AF773226

Physics of Atoms and Molecules

Physics of Atoms and Molecules

Edited by
Mark Romano

Willford Press
www.willfordpress.com

Published by Willford Press,
118-35 Queens Blvd., Suite 400,
Forest Hills, NY 11375, USA

Copyright © 2018 Willford Press

This book contains information obtained from authentic and highly regarded sources. All chapters are published with permission under the Creative Commons Attribution Share Alike License or equivalent. A wide variety of references are listed. Permissions and sources are indicated; for detailed attributions, please refer to the permissions page. Reasonable efforts have been made to publish reliable data and information, but the authors, editors and publisher cannot assume any responsibility for the vailidity of all materials or the consequences of their use.

Trademark Notice: Registered trademark of products or corporate names are used only for explanation and identification without intent to infringe.

ISBN: 978-1-68285-498-3

Cataloging-in-Publication Data

Physics of atoms and molecules / edited by Mark Romano.
p. cm.
Includes bibliographical references and index.
ISBN 978-1-68285-498-3
1. Atoms. 2. Molecules. 3. Atom-molecule collisions. 4. Physics. I. Romano, Mark.
QC171.2 .P49 2018
539.7--dc23

For information on all Willford Press publications
visit our website at www.willfordpress.com

WILLFORD PRESS

Printed in China.

Contents

Preface

Atomic physics is a sub-branch of physics which deals with the study of atoms, their nucleus and the electrons surrounding it. It is used to expand fields like nuclear weapons and nuclear power. Similarly, molecular physics refers to the study of molecular dynamics, the characteristics of molecules and also the chemical bonds between atoms. The other fields which use the elements of atomic and molecular physics are chemical physics, theoretical chemistry, physical chemistry, etc. This book aims to shed light on some of the unexplored aspects of the physics of atoms and molecules. Some of the diverse topics covered in it address the varied branches that fall under this category. This book is appropriate for those seeking detailed information in this area.

A detailed account of the significant topics covered in this book is provided below:

Chapter 1- An atom is the smallest particles of matter. It is composed of a nucleus and electron which is connected to the nucleus of the atom. Two or more atoms bond together form a molecule. This chapter will provide an integrated understanding of atomic and molecular physics.

Chapter 2- Classical mechanics could not explain many experiments such as the photoelectric effect, Compton effect and proposals by Bohr and Rutherford. It was only with the development of quantum mechanics that many of these theories could be satisfactorily understood. This chapter focuses on Maxwell's equations, electromagnetic radiation, wave-particle duality, radiative transition, spectral broadening, etc.

Chapter 3- Equating a hydrogen-like atom with the Schrödinger equation is an important experiment. A hydrogen-like atom has a nucleus and one electron. The major components of the atomic structure of hydrogen are discussed in this chapter.

Chapter 4- Chemical bonding, ionic bonding and other processes can be better understood with the help of quantum mechanics. Molecular formation and Coulomb force are one of the significant and important topics related to molecular physics. The following chapter unfolds its crucial aspects in a critical yet systematic manner.

It gives me an immense pleasure to thank our entire team for their efforts. Finally in the end, I would like to thank my family and colleagues who have been a great source of inspiration and support.

Editor

1

A Brief Introduction of Atoms and Molecules

An atom is the smallest particles of matter. It is composed of a nucleus and electron which is connected to the nucleus of the atom. Two or more atoms bond together form a molecule. This chapter will provide an integrated understanding of atomic and molecular physics.

Atom

An atom is the smallest constituent unit of ordinary matter that has the properties of a chemical element. Every solid, liquid, gas, and plasma is composed of neutral or ionized atoms. Atoms are very small; typical sizes are around 100 picometers (a ten-billionth of a meter, in the short scale).

Atoms are small enough that attempting to predict their behavior using classical physics - as if they were billiard balls, for example - gives noticeably incorrect predictions due to quantum effects. Through the development of physics, atomic models have incorporated quantum principles to better explain and predict the behavior.

Every atom is composed of a nucleus and one or more electrons bound to the nucleus. The nucleus is made of one or more protons and typically a similar number of neutrons. Protons and neutrons are called nucleons. More than 99.94% of an atom's mass is in the nucleus. The protons have a positive electric charge, the electrons have a negative electric charge, and the neutrons have no electric charge. If the number of protons and electrons are equal, that atom is electrically neutral. If an atom has more or fewer electrons than protons, then it has an overall negative or positive charge, respectively, and it is called an ion.

The electrons of an atom are attracted to the protons in an atomic nucleus by this electromagnetic force. The protons and neutrons in the nucleus are attracted to each other by a different force, the nuclear force, which is usually stronger than the electromagnetic force repelling the positively charged protons from one another. Under certain circumstances the repelling electromagnetic force becomes stronger than the nuclear force, and nucleons can be ejected from the nucleus, leaving behind a different element: nuclear decay resulting in nuclear transmutation.

The number of protons in the nucleus defines to what chemical element the atom belongs: for example, all copper atoms contain 29 protons. The number of neutrons defines the isotope of the element. The number of electrons influences the magnetic properties of an atom. Atoms can attach to one or more other atoms by chemical bonds to form chemical compounds such as molecules. The ability of atoms to associate and dissociate is responsible for most of the physical changes observed in nature, and is the subject of the discipline of chemistry.

History of Atomic Theory

Atoms in Philosophy

The idea that matter is made up of discrete units is a very old idea, appearing in many ancient cultures such as Greece and India. The word "atom" was coined by ancient Greek philosophers. However, these ideas were founded in philosophical and theological reasoning rather than evidence and experimentation. As a result, their views on what atoms look like and how they behave were incorrect. They also could not convince everybody, so atomism was but one of a number of competing theories on the nature of matter. It was not until the 19th century that the idea was embraced and refined by scientists, when the blossoming science of chemistry produced discoveries that only the concept of atoms could explain.

First Evidence-based Theory

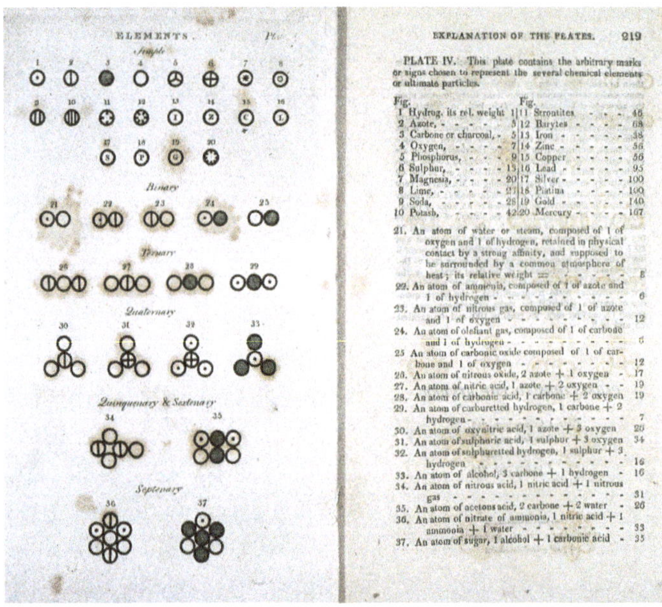

Various atoms and molecules as depicted in John Dalton's *A New System of Chemical Philosophy* (1808).

In the early 1800s, John Dalton used the concept of atoms to explain why elements always react in ratios of small whole numbers (the law of multiple proportions). For instance, there are two types of tin oxide: one is 88.1% tin and 11.9% oxygen and the other is 78.7% tin and 21.3% oxygen (tin(II) oxide and tin dioxide respectively). This means that 100g of tin will combine either with 13.5g or 27g of oxygen. 13.5 and 27 form a ratio of 1:2, a ratio of small whole numbers. This common pattern in chemistry suggested to Dalton that elements react in whole number multiples of discrete units—in other words, atoms. In the case of tin oxides, one tin atom will combine with either one or two oxygen atoms.

Dalton also believed atomic theory could explain why water absorbs different gases in different proportions. For example, he found that water absorbs carbon dioxide far better than it absorbs nitrogen. Dalton hypothesized this was due to the differences between the masses and configurations of the gases' respective particles, and carbon dioxide molecules (CO_2) are heavier and larger than nitrogen molecules (N_2).

Brownian Motion

In 1827, botanist Robert Brown used a microscope to look at dust grains floating in water and discovered that they moved about erratically, a phenomenon that became known as "Brownian motion". This was thought to be caused by water molecules knocking the grains about. In 1905, Albert Einstein proved the reality of these molecules and their motions by producing the first Statistical physics analysis of Brownian motion. French physicist Jean Perrin used Einstein's work to experimentally determine the mass and dimensions of atoms, thereby conclusively verifying Dalton's atomic theory.

Discovery of the Electron

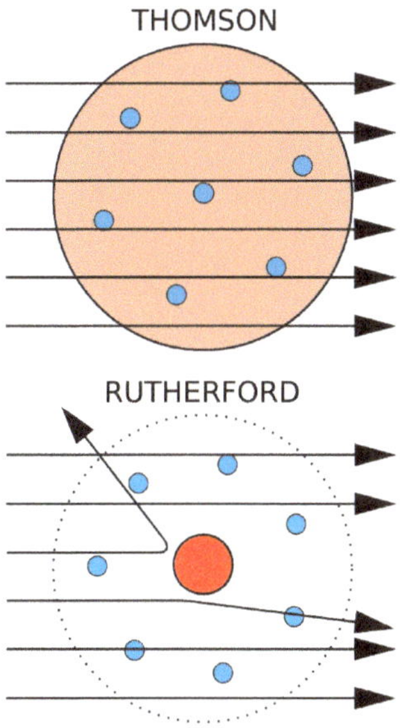

The Geiger–Marsden experiment *Top:* Expected results: alpha particles passing through the plum pudding model of the atom with negligible deflection. *Bottom:* Observed results: a small portion of the particles were deflected by the concentrated positive charge of the nucleus.

The physicist J. J. Thomson measured the mass of cathode rays, showing they were made of particles, but were around 1800 times lighter than the lightest atom, hydrogen. Therefore, they were not atoms, but a new particle, the first *subatomic* particle to be discovered, which he originally called "*corpuscle*" but was later named *electron*, after particles postulated by George Johnstone Stoney in 1874. He also showed they were identical to particles given off by photoelectric and radioactive materials. It was quickly recognized that they are the particles that carry electric currents in metal wires, and carry the negative electric charge within atoms. Thomson was given the 1906 Nobel Prize in Physics for this work. Thus he overturned the belief that atoms are the indivisible, ultimate particles of matter. Thomson also incorrectly postulated that the low mass, negatively charged electrons were distributed throughout the atom in a uniform sea of positive charge. This became known as the plum pudding model.

Discovery of the Nucleus

In 1909, Hans Geiger and Ernest Marsden, under the direction of Ernest Rutherford, bombarded a metal foil with alpha particles to observe how they scattered. They expected all the alpha particles to pass straight through with little deflection, because Thomson's model said that the charges in the atom are so diffuse that their electric fields could not affect the alpha particles much. However, Geiger and Marsden spotted alpha particles being deflected by angles greater than 90°, which was supposed to be impossible according to Thomson's model. To explain this, Rutherford proposed that the positive charge of the atom is concentrated in a tiny nucleus at the center of the atom. Rutherford compared his findings to firing a 15-inch shell at a sheet of tissue paper and it coming back to hit you.

Discovery of Isotopes

While experimenting with the products of radioactive decay, in 1913 radiochemist Frederick Soddy discovered that there appeared to be more than one type of atom at each position on the periodic table. The term isotope was coined by Margaret Todd as a suitable name for different atoms that belong to the same element. J.J. Thomson created a technique for separating atom types through his work on ionized gases, which subsequently led to the discovery of stable isotopes.

Bohr Model

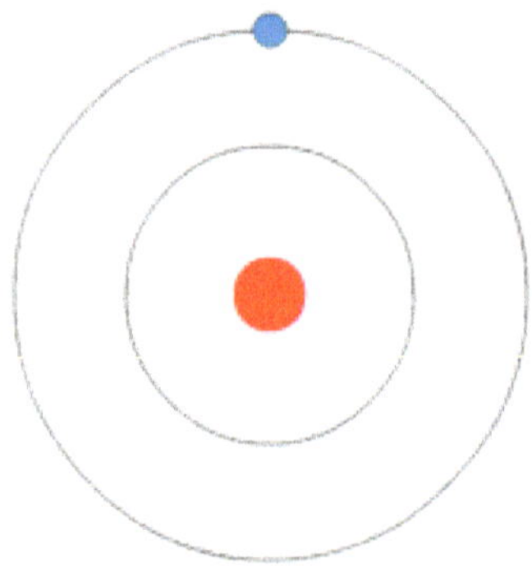

The Bohr model of the atom, with an electron making instantaneous "quantum leaps" from one orbit to another. This model is obsolete.

In 1913 the physicist Niels Bohr proposed a model in which the electrons of an atom were assumed to orbit the nucleus but could only do so in a finite set of orbits, and could jump between these orbits only in discrete changes of energy corresponding to absorption or radiation of a photon. This quantization was used to explain why the electrons orbits are stable (given that normally, charges in acceleration, including circular motion, lose kinetic energy which is emitted as electromagnetic radiation) and why elements absorb and emit electromagnetic radiation in discrete spectra.

Later in the same year Henry Moseley provided additional experimental evidence in favor of Niels Bohr's theory. These results refined Ernest Rutherford's and Antonius Van den Broek's model, which proposed that the atom contains in its nucleus a number of positive nuclear charges that is equal to its (atomic) number in the periodic table. Until these experiments, atomic number was not known to be a physical and experimental quantity. That it is equal to the atomic nuclear charge remains the accepted atomic model today.

Chemical Bonding Explained

Chemical bonds between atoms were now explained, by Gilbert Newton Lewis in 1916, as the interactions between their constituent electrons. As the chemical properties of the elements were known to largely repeat themselves according to the periodic law, in 1919 the American chemist Irving Langmuir suggested that this could be explained if the electrons in an atom were connected or clustered in some manner. Groups of electrons were thought to occupy a set of electron shells about the nucleus.

Further Developments in Quantum Physics

The Stern–Gerlach experiment of 1922 provided further evidence of the quantum nature of the atom. When a beam of silver atoms was passed through a specially shaped magnetic field, the beam was split based on the direction of an atom's angular momentum, or spin. As this direction is random, the beam could be expected to spread into a line. Instead, the beam was split into two parts, depending on whether the atomic spin was oriented up or down.

In 1924, Louis de Broglie proposed that all particles behave to an extent like waves. In 1926, Erwin Schrödinger used this idea to develop a mathematical model of the atom that described the electrons as three-dimensional waveforms rather than point particles. A consequence of using waveforms to describe particles is that it is mathematically impossible to obtain precise values for both the position and momentum of a particle at a given point in time; this became known as the uncertainty principle, formulated by Werner Heisenberg in 1926. In this concept, for a given accuracy in measuring a position one could only obtain a range of probable values for momentum, and vice versa. This model was able to explain observations of atomic behavior that previous models could not, such as certain structural and spectral patterns of atoms larger than hydrogen. Thus, the planetary model of the atom was discarded in favor of one that described atomic orbital zones around the nucleus where a given electron is most likely to be observed.

Discovery of the Neutron

The development of the mass spectrometer allowed the mass of atoms to be measured with increased accuracy. The device uses a magnet to bend the trajectory of a beam of ions, and the amount of deflection is determined by the ratio of an atom's mass to its charge. The chemist Francis William Aston used this instrument to show that isotopes had different masses. The atomic mass of these isotopes varied by integer amounts, called the whole number rule. The explanation for these different isotopes awaited the discovery of the neutron, an uncharged particle with a mass similar to the proton, by the physicist James Chadwick in 1932. Isotopes were then explained as elements with the same number of protons, but different numbers of neutrons within the nucleus.

Fission, High-energy Physics and Condensed Matter

In 1938, the German chemist Otto Hahn, a student of Rutherford, directed neutrons onto uranium atoms expecting to get transuranium elements. Instead, his chemical experiments showed barium as a product. A year later, Lise Meitner and her nephew Otto Frisch verified that Hahn's result were the first experimental *nuclear fission*. In 1944, Hahn received the Nobel prize in chemistry. Despite Hahn's efforts, the contributions of Meitner and Frisch were not recognized.

In the 1950s, the development of improved particle accelerators and particle detectors allowed scientists to study the impacts of atoms moving at high energies. Neutrons and protons were found to be hadrons, or composites of smaller particles called quarks. The standard model of particle physics was developed that so far has successfully explained the properties of the nucleus in terms of these sub-atomic particles and the forces that govern their interactions.

Structure

Subatomic Particles

Though the word *atom* originally denoted a particle that cannot be cut into smaller particles, in modern scientific usage the atom is composed of various subatomic particles. The constituent particles of an atom are the electron, the proton and the neutron; all three are fermions. However, the hydrogen-1 atom has no neutrons and the hydron ion has no electrons.

The electron is by far the least massive of these particles at 9.11×10^{-31} kg, with a negative electrical charge and a size that is too small to be measured using available techniques. It is the lightest particle with a positive rest mass measured. Under ordinary conditions, electrons are bound to the positively charged nucleus by the attraction created from opposite electric charges. If an atom has more or fewer electrons than its atomic number, then it becomes respectively negatively or positively charged as a whole; a charged atom is called an ion. Electrons have been known since the late 19th century, mostly thanks to J.J. Thomson.

Protons have a positive charge and a mass 1,836 times that of the electron, at 1.6726×10^{-27} kg. The number of protons in an atom is called its atomic number. Ernest Rutherford (1919) observed that nitrogen under alpha-particle bombardment ejects what appeared to be hydrogen nuclei. By 1920 he had accepted that the hydrogen nucleus is a distinct particle within the atom and named it proton.

Neutrons have no electrical charge and have a free mass of 1,839 times the mass of the electron, or 1.6929×10^{-27} kg, the heaviest of the three constituent particles, but it can be reduced by the nuclear binding energy. Neutrons and protons (collectively known as nucleons) have comparable dimensions—on the order of 2.5×10^{-15} m—although the ‹surface› of these particles is not sharply defined. The neutron was discovered in 1932 by the English physicist James Chadwick.

In the Standard Model of physics, electrons are truly elementary particles with no internal structure. However, both protons and neutrons are composite particles composed of elementary particles called quarks. There are two types of quarks in atoms, each having a fractional electric charge. Protons are composed of two up quarks (each with charge +2/3) and one down quark (with a charge of −1/3). Neutrons consist of one up quark and two down quarks. This distinction accounts for the difference in mass and charge between the two particles.

The quarks are held together by the strong interaction (or strong force), which is mediated by gluons. The protons and neutrons, in turn, are held to each other in the nucleus by the nuclear force, which is a residuum of the strong force that has somewhat different range-properties. The gluon is a member of the family of gauge bosons, which are elementary particles that mediate physical forces.

Nucleus

All the bound protons and neutrons in an atom make up a tiny atomic nucleus, and are collectively called nucleons. The radius of a nucleus is approximately equal to $1.07\sqrt[3]{A}$ fm, where A is the total number of nucleons. This is much smaller than the radius of the atom, which is on the order of 10^5 fm. The nucleons are bound together by a short-ranged attractive potential called the residual strong force. At distances smaller than 2.5 fm this force is much more powerful than the electrostatic force that causes positively charged protons to repel each other.

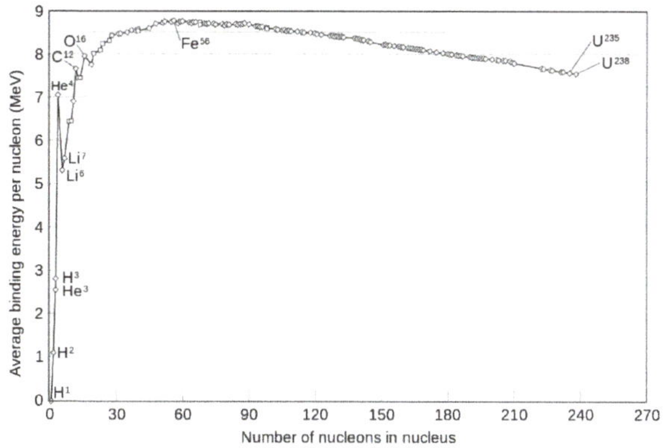

The binding energy needed for a nucleon to escape the nucleus, for various isotopes

Atoms of the same element have the same number of protons, called the atomic number. Within a single element, the number of neutrons may vary, determining the isotope of that element. The total number of protons and neutrons determine the nuclide. The number of neutrons relative to the protons determines the stability of the nucleus, with certain isotopes undergoing radioactive decay.

The proton, the electron, and the neutron are classified as fermions. Fermions obey the Pauli exclusion principle which prohibits *identical* fermions, such as multiple protons, from occupying the same quantum state at the same time. Thus, every proton in the nucleus must occupy a quantum state different from all other protons, and the same applies to all neutrons of the nucleus and to all electrons of the electron cloud. However, a proton and a neutron are allowed to occupy the same quantum state.

For atoms with low atomic numbers, a nucleus that has more neutrons than protons tends to drop to a lower energy state through radioactive decay so that the neutron–proton ratio is closer to one. However, as the atomic number increases, a higher proportion of neutrons is required to offset the mutual repulsion of the protons. Thus, there are no stable nuclei with equal proton and neutron numbers above atomic number $Z = 20$ (calcium) and as Z increases, the neutron–proton ratio of stable isotopes increases. The stable isotope with the highest proton–neutron ratio is lead-208 (about 1.5).

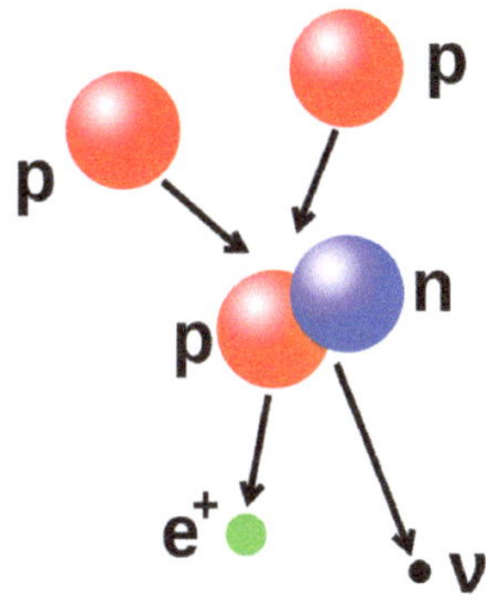

Illustration of a nuclear fusion process that forms a deuterium nucleus, consisting of a proton and a neutron, from two protons. A positron (e^+)—an antimatter electron—is emitted along with an electron neutrino

The number of protons and neutrons in the atomic nucleus can be modified, although this can require very high energies because of the strong force. Nuclear fusion occurs when multiple atomic particles join to form a heavier nucleus, such as through the energetic collision of two nuclei. For example, at the core of the Sun protons require energies of 3–10 keV to overcome their mutual repulsion—the coulomb barrier—and fuse together into a single nucleus. Nuclear fission is the opposite process, causing a nucleus to split into two smaller nuclei—usually through radioactive decay. The nucleus can also be modified through bombardment by high energy subatomic particles or photons. If this modifies the number of protons in a nucleus, the atom changes to a different chemical element.

If the mass of the nucleus following a fusion reaction is less than the sum of the masses of the separate particles, then the difference between these two values can be emitted as a type of usable energy (such as a gamma ray, or the kinetic energy of a beta particle), as described by Albert Einstein's mass–energy equivalence formula, $E = mc^2$, where m is the mass loss and c is the speed of light. This deficit is part of the binding energy of the new nucleus, and it is the non-recoverable loss of the energy that causes the fused particles to remain together in a state that requires this energy to separate.

The fusion of two nuclei that create larger nuclei with lower atomic numbers than iron and nickel—a total nucleon number of about 60—is usually an exothermic process that releases more energy than is required to bring them together. It is this energy-releasing process that makes nuclear fusion in stars a self-sustaining reaction. For heavier nuclei, the binding energy per nucleon in the nucleus begins to decrease. That means fusion processes producing nuclei that have atomic numbers higher than about 26, and atomic masses higher than about 60, is an endothermic process. These more massive nuclei can not undergo an energy-producing fusion reaction that can sustain the hydrostatic equilibrium of a star.

Electron Cloud

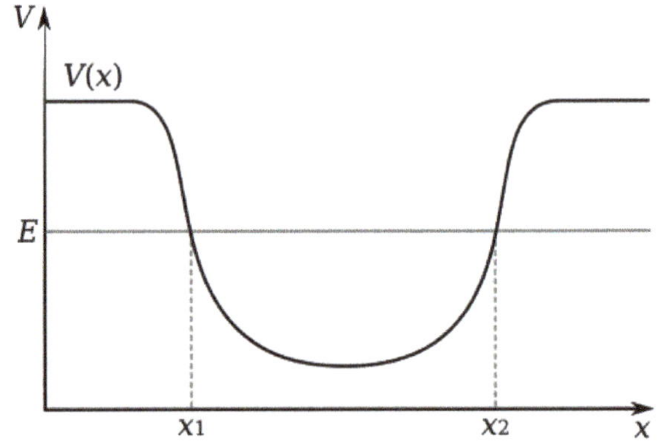

A potential well, showing, according to classical mechanics, the minimum energy $V(x)$ needed to reach each position x. Classically, a particle with energy E is constrained to a range of positions between x_1 and x_2

The electrons in an atom are attracted to the protons in the nucleus by the electromagnetic force. This force binds the electrons inside an electrostatic potential well surrounding the smaller nucleus, which means that an external source of energy is needed for the electron to escape. The closer an electron is to the nucleus, the greater the attractive force. Hence electrons bound near the center of the potential well require more energy to escape than those at greater separations.

Electrons, like other particles, have properties of both a particle and a wave. The electron cloud is a region inside the potential well where each electron forms a type of three-dimensional standing wave—a wave form that does not move relative to the nucleus. This behavior is defined by an atomic orbital, a mathematical function that characterises the probability that an electron appears to be at a particular location when its position is measured. Only a discrete (or quantized) set of these orbitals exist around the nucleus, as other possible wave patterns rapidly decay into a more stable form. Orbitals can have one or more ring or node structures, and differ from each other in size, shape and orientation.

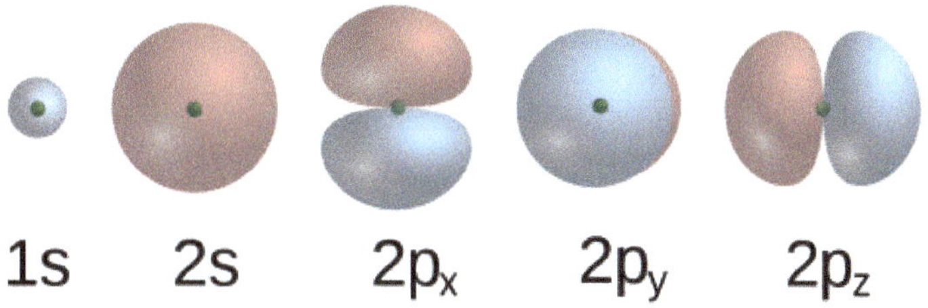

Wave functions of the first five atomic orbitals. The three 2p orbitals each display a single angular node that has an orientation and a minimum at the center

Each atomic orbital corresponds to a particular energy level of the electron. The electron can change its state to a higher energy level by absorbing a photon with sufficient energy to boost it into the new quantum state. Likewise, through spontaneous emission, an electron in a higher energy state can drop to a lower energy state while radiating the excess energy as a photon. These characteristic energy values, defined by the differences in the energies of the quantum states, are responsible for atomic spectral lines.

The amount of energy needed to remove or add an electron—the electron binding energy—is far less than the binding energy of nucleons. For example, it requires only 13.6 eV to strip a ground-state electron from a hydrogen atom, compared to 2.23 *million* eV for splitting a deuterium nucleus. Atoms are electrically neutral if they have an equal number of protons and electrons. Atoms that have either a deficit or a surplus of electrons are called ions. Electrons that are farthest from the nucleus may be transferred to other nearby atoms or shared between atoms. By this mechanism, atoms are able to bond into molecules and other types of chemical compounds like ionic and covalent network crystals.

Properties

Nuclear Properties

By definition, any two atoms with an identical number of *protons* in their nuclei belong to the same chemical element. Atoms with equal numbers of protons but a different number of *neutrons* are different isotopes of the same element. For example, all hydrogen atoms admit exactly one proton, but isotopes exist with no neutrons (hydrogen-1, by far the most common form, also called protium), one neutron (deuterium), two neutrons (tritium) and more than two neutrons. The known elements form a set of atomic numbers, from the single proton element hydrogen up to the 118-proton element oganesson. All known isotopes of elements with atomic numbers greater than 82 are radioactive, although the radioactivity of element 83 (bismuth) is so slight as to be practically negligible.

About 339 nuclides occur naturally on Earth, of which 254 (about 75%) have not been observed to decay, and are referred to as "stable isotopes". However, only 90 of these nuclides are stable to all decay, even in theory. Another 164 (bringing the total to 254) have not been observed to decay, even though in theory it is energetically possible. These are also formally classified as "stable". An additional 34 radioactive nuclides have half-lives longer than 80 million years, and are long-lived enough to be present from the birth of the solar system. This collection of 288 nuclides are known as primordial nuclides. Finally, an additional 51 short-lived nuclides are known to occur naturally, as daughter products of primordial nuclide decay (such as radium from uranium), or else as products of natural energetic processes on Earth, such as cosmic ray bombardment (for example, carbon-14).

For 80 of the chemical elements, at least one stable isotope exists. As a rule, there is only a handful of stable isotopes for each of these elements, the average being 3.2 stable isotopes per element. Twenty-six elements have only a single stable isotope, while the largest number of stable isotopes observed for any element is ten, for the element tin. Elements 43, 61, and all elements numbered 83 or higher have no stable isotopes.

Stability of isotopes is affected by the ratio of protons to neutrons, and also by the presence of certain "magic numbers" of neutrons or protons that represent closed and filled quantum shells. These quantum shells correspond to a set of energy levels within the shell model of the nucleus; filled shells, such as the filled shell of 50 protons for tin, confers unusual stability on the nuclide. Of the 254 known stable nuclides, only four have both an odd number of protons *and* odd number of neutrons: hydrogen-2 (deuterium), lithium-6, boron-10 and nitrogen-14. Also, only four naturally occurring, radioactive odd–odd nuclides have a half-life over a billion years: potassium-40, vanadium-50, lanthanum-138 and tantalum-180m. Most odd–odd nuclei are highly unstable with respect to beta decay, because the decay products are even–even, and are therefore more strongly bound, due to nuclear pairing effects.

Mass

The large majority of an atom's mass comes from the protons and neutrons that make it up. The total number of these particles (called "nucleons") in a given atom is called the mass number. It is a positive integer and dimensionless (instead of having dimension of mass), because it expresses a count. An example of use of a mass number is "carbon-12," which has 12 nucleons (six protons and six neutrons).

The actual mass of an atom at rest is often expressed using the unified atomic mass unit (u), also called dalton (Da). This unit is defined as a twelfth of the mass of a free neutral atom of carbon-12, which is approximately 1.66×10^{-27} kg. Hydrogen-1 (the lightest isotope of hydrogen which is also the nuclide with the lowest mass) has an atomic weight of 1.007825 u. The value of this number is called the atomic mass. A given atom has an atomic mass approximately equal (within 1%) to its mass number times the atomic mass unit (for example the mass of a nitrogen-14 is roughly 14 u). However, this number will not be exactly an integer except in the case of carbon-12. The heaviest stable atom is lead-208, with a mass of 207.9766521 u.

As even the most massive atoms are far too light to work with directly, chemists instead use the unit of moles. One mole of atoms of any element always has the same number of atoms (about 6.022×10^{23}). This number was chosen so that if an element has an atomic mass of 1 u, a mole of atoms of that element has a mass close to one gram. Because of the definition of the unified atomic mass unit, each carbon-12 atom has an atomic mass of exactly 12 u, and so a mole of carbon-12 atoms weighs exactly 0.012 kg.

Shape and Size

Atoms lack a well-defined outer boundary, so their dimensions are usually described in terms of an atomic radius. This is a measure of the distance out to which the electron cloud extends from the nucleus. However, this assumes the atom to exhibit a spherical shape, which is only obeyed for atoms in vacuum or free space. Atomic radii may be derived from the distances between two nuclei when the two atoms are joined in a chemical bond. The radius varies with the location of an atom on the atomic chart, the type of chemical bond, the number of neighboring atoms (coordination number) and a quantum mechanical property known as spin. On the periodic table of the elements, atom size tends to increase when moving down columns, but decrease when moving across rows (left to right). Consequently, the smallest atom is helium with a radius of 32 pm, while one of the largest is caesium at 225 pm.

When subjected to external forces, like electrical fields, the shape of an atom may deviate from spherical symmetry. The deformation depends on the field magnitude and the orbital type of outer shell electrons, as shown by group-theoretical considerations. Aspherical deviations might be elicited for instance in crystals, where large crystal-electrical fields may occur at low-symmetry lattice sites. Significant ellipsoidal deformations have recently been shown to occur for sulfur ions and chalcogen ions in pyrite-type compounds.

Atomic dimensions are thousands of times smaller than the wavelengths of light (400–700 nm) so they cannot be viewed using an optical microscope. However, individual atoms can be observed using a scanning tunneling microscope. To visualize the minuteness of the atom, consider that a typical human hair is about 1 million carbon atoms in width. A single drop of water contains about 2 sextillion (2×10^{21}) atoms of oxygen, and twice the number of hydrogen atoms. A single carat diamond with a mass of 2×10^{-4} kg contains about 10 sextillion (10^{22}) atoms of carbon. If an apple were magnified to the size of the Earth, then the atoms in the apple would be approximately the size of the original apple.

Radioactive Decay

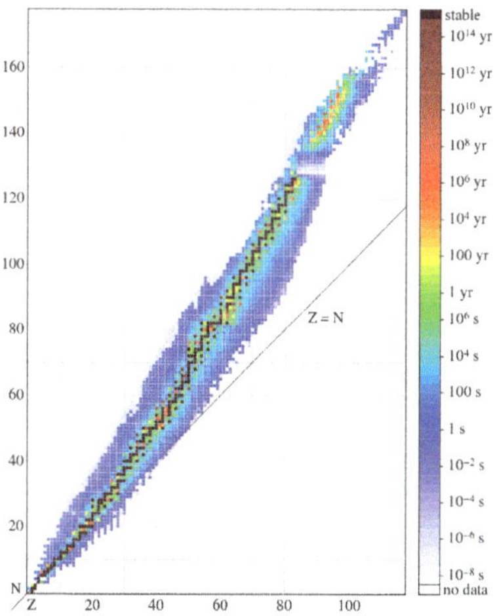

This diagram shows the half-life ($T_{½}$) of various isotopes with Z protons and N neutrons

Every element has one or more isotopes that have unstable nuclei that are subject to radioactive decay, causing the nucleus to emit particles or electromagnetic radiation. Radioactivity can occur when the radius of a nucleus is large compared with the radius of the strong force, which only acts over distances on the order of 1 fm.

The most common forms of radioactive decay are:

- Alpha decay: this process is caused when the nucleus emits an alpha particle, which is a helium nucleus consisting of two protons and two neutrons. The result of the emission is a new element with a lower atomic number.
- Beta decay (and electron capture): these processes are regulated by the weak force, and result from a transformation of a neutron into a proton, or a proton into a neutron. The neutron to proton transition is accompanied by the emission of an electron and an antineutrino, while proton to neutron transition (except in electron capture) causes the emission of a positron and a neutrino. The electron or positron emissions are called beta particles. Beta decay either increases or decreases the atomic number of the nucleus by one. Electron capture is more common than positron emission, because it requires less energy. In this type of decay, an electron is absorbed by the nucleus, rather than a positron emitted from the nucleus. A neutrino is still emitted in this process, and a proton changes to a neutron.
- Gamma decay: this process results from a change in the energy level of the nucleus to a lower state, resulting in the emission of electromagnetic radiation. The excited state of a nucleus which results in gamma emission usually occurs following the emission of an alpha or a beta particle. Thus, gamma decay usually follows alpha or beta decay.

Other more rare types of radioactive decay include ejection of neutrons or protons or clusters of nucleons from a nucleus, or more than one beta particle. An analog of gamma emission which allows excited nuclei to lose energy in a different way, is internal conversion— a process that produces high-speed electrons that are not beta rays, followed by production of high-energy photons that are not gamma rays. A few large nuclei explode into two or more charged fragments of varying masses plus several neutrons, in a decay called spontaneous nuclear fission.

Each radioactive isotope has a characteristic decay time period—the half-life—that is determined by the amount of time needed for half of a sample to decay. This is an exponential decay process that steadily decreases the proportion of the remaining isotope by 50% every half-life. Hence after two half-lives have passed only 25% of the isotope is present, and so forth.

Magnetic Moment

Elementary particles possess an intrinsic quantum mechanical property known as spin. This is analogous to the angular momentum of an object that is spinning around its center of mass, although strictly speaking these particles are believed to be point-like and cannot be said to be rotating. Spin is measured in units of the reduced Planck constant (ħ), with electrons, protons and neutrons all having spin ½ ħ, or «spin-½». In an atom, electrons in motion around the nucleus possess orbital angular momentum in addition to their spin, while the nucleus itself possesses angular momentum due to its nuclear spin.

The magnetic field produced by an atom—its magnetic moment—is determined by these various forms of angular momentum, just as a rotating charged object classically produces a magnetic field. However, the most dominant contribution comes from electron spin. Due to the nature of electrons to obey the Pauli exclusion principle, in which no two electrons may be found in the same quantum state, bound electrons pair up with each other, with one member of each pair in a spin up state and the other in the opposite, spin down state. Thus these spins cancel each other out, reducing the total magnetic dipole moment to zero in some atoms with even number of electrons.

In ferromagnetic elements such as iron, cobalt and nickel, an odd number of electrons leads to an unpaired electron and a net overall magnetic moment. The orbitals of neighboring atoms overlap and a lower energy state is achieved when the spins of unpaired electrons are aligned with each other, a spontaneous process known as an exchange interaction. When the magnetic moments of ferromagnetic atoms are lined up, the material can produce a measurable macroscopic field. Paramagnetic materials have atoms with magnetic moments that line up in random directions when no magnetic field is present, but the magnetic moments of the individual atoms line up in the presence of a field.

The nucleus of an atom will have no spin when it has even numbers of both neutrons and protons, but for other cases of odd numbers, the nucleus may have a spin. Normally nuclei with spin are aligned in random directions because of thermal equilibrium. However, for certain elements (such as xenon-129) it is possible to polarize a significant proportion of the nuclear spin states so that they are aligned in the same direction—a condition called hyperpolarization. This has important applications in magnetic resonance imaging.

Energy Levels

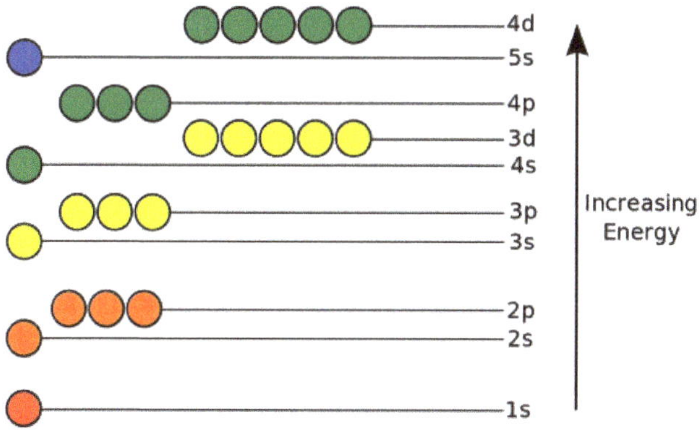

These electron's energy levels (not to scale) are sufficient for ground states of atoms up to cadmium ($5s^2$ $4d^{10}$) inclusively. Do not forget that even the top of the diagram is lower than an unbound electron state

The potential energy of an electron in an atom is negative, its dependence of its position reaches the minimum (the most absolute value) inside the nucleus, and vanishes when the distance from the nucleus goes to infinity, roughly in an inverse proportion to the distance. In the quantum-mechanical model, a bound electron can only occupy a set of states centered on the nucleus, and each state corresponds to a specific energy level; time-independent Schrödinger equation for theoretical explanation. An energy level can be measured by the amount of energy needed to unbind the electron from the atom, and is usually given in units of electronvolts (eV). The lowest energy state of a bound electron is called the ground state, i.e. stationary state, while an electron transition to a higher level results in an excited state. The electron's energy raises when n increases because the (average) distance to the nucleus increases. Dependence of the energy on ℓ is caused not by electrostatic potential of the nucleus, but by interaction between electrons.

For an electron to transition between two different states, e.g. grounded state to first excited level (ionization), it must absorb or emit a photon at an energy matching the difference in the potential energy of those levels, according to Niels Bohr model, what can be precisely calculated by the Schrödinger equation. Electrons jump between orbitals in a particle-like fashion. For example, if a single photon strikes the electrons, only a single electron changes states in response to the photon.

The energy of an emitted photon is proportional to its frequency, so these specific energy levels appear as distinct bands in the electromagnetic spectrum. Each element has a characteristic spectrum that can depend on the nuclear charge, subshells filled by electrons, the electromagnetic interactions between the electrons and other factors.

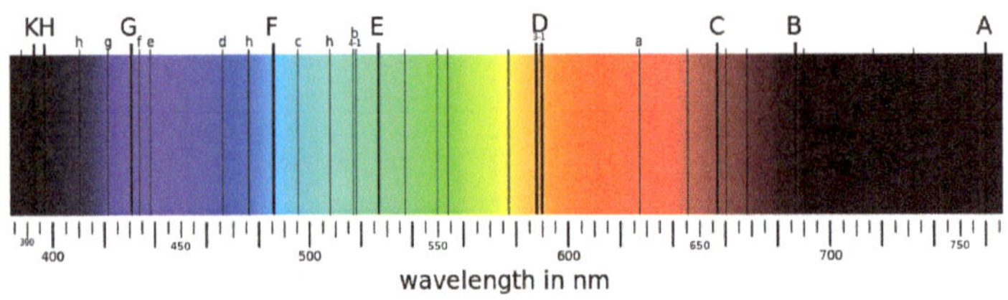

An example of absorption lines in a spectrum

When a continuous spectrum of energy is passed through a gas or plasma, some of the photons are absorbed by atoms, causing electrons to change their energy level. Those excited electrons that remain bound to their atom spontaneously emit this energy as a photon, traveling in a random direction, and so drop back to lower energy levels. Thus the atoms behave like a filter that forms a series of dark absorption bands in the energy output. (An observer viewing the atoms from a view that does not include the continuous spectrum in the background, instead sees a series of emission lines from the photons emitted by the atoms.) Spectroscopic measurements of the strength and width of atomic spectral lines allow the composition and physical properties of a substance to be determined.

Close examination of the spectral lines reveals that some display a fine structure splitting. This occurs because of spin–orbit coupling, which is an interaction between the spin and motion of the outermost electron. When an atom is in an external magnetic field, spectral lines become split into three or more components; a phenomenon called the Zeeman effect. This is caused by the interaction of the magnetic field with the magnetic moment of the atom and its electrons. Some atoms can have multiple electron configurations with the same energy level, which thus appear as a single spectral line. The interaction of the magnetic field with the atom shifts these electron configurations to slightly different energy levels, resulting in multiple spectral lines. The presence of an external electric field can cause a comparable splitting and shifting of spectral lines by modifying the electron energy levels, a phenomenon called the Stark effect.

If a bound electron is in an excited state, an interacting photon with the proper energy can cause stimulated emission of a photon with a matching energy level. For this to occur, the electron must drop to a lower energy state that has an energy difference matching the energy of the interacting photon. The emitted photon and the interacting photon then move off in parallel and with matching phases. That is, the wave patterns of the two photons are synchronized. This physical property is used to make lasers, which can emit a coherent beam of light energy in a narrow frequency band.

Valence and Bonding Behavior

Valency is the combining power of an element. It is equal to number of hydrogen atoms that atom can combine or displace in forming compounds. The outermost electron shell of an atom in its uncombined state is known as the valence shell, and the electrons in that shell are called valence electrons. The number of valence electrons determines the bonding behavior with other atoms. Atoms tend to chemically react with each other in a manner that fills (or empties) their outer valence shells. For example, a transfer of a single electron between atoms is a useful approximation for bonds that form between atoms with one-electron more than a filled shell, and others that are one-electron short of a full shell, such as occurs in the compound sodium chloride and other chemical ionic salts. However, many elements display multiple valences, or tendencies to share differing numbers of electrons in different compounds. Thus, chemical bonding between these elements takes many forms of electron-sharing that are more than simple electron transfers. Examples include the element carbon and the organic compounds.

The chemical elements are often displayed in a periodic table that is laid out to display recurring chemical properties, and elements with the same number of valence electrons form a group that is aligned in the same column of the table. (The horizontal rows correspond to the filling of a quantum shell of electrons.) The elements at the far right of the table have their outer shell completely filled with electrons, which results in chemically inert elements known as the noble gases.

States

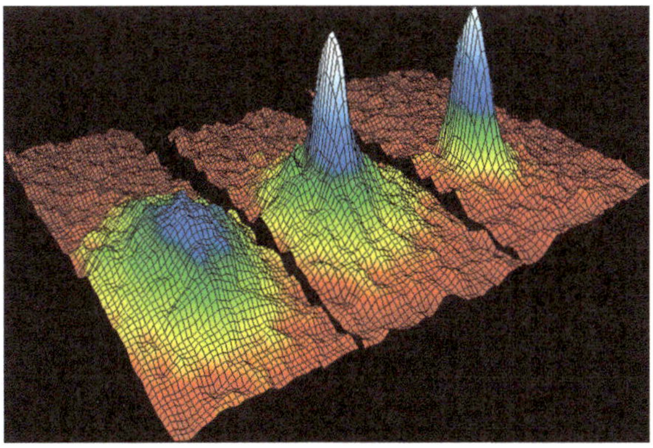

Snapshots illustrating the formation of a Bose–Einstein condensate

Quantities of atoms are found in different states of matter that depend on the physical conditions, such as temperature and pressure. By varying the conditions, materials can transition between solids, liquids, gases and plasmas. Within a state, a material can also exist in different allotropes. An example of this is solid carbon, which can exist as graphite or diamond. Gaseous allotropes exist as well, such as dioxygen and ozone.

At temperatures close to absolute zero, atoms can form a Bose–Einstein condensate, at which point quantum mechanical effects, which are normally only observed at the atomic scale, become apparent on a macroscopic scale. This super-cooled collection of atoms then behaves as a single super atom, which may allow fundamental checks of quantum mechanical behavior.

Identification

Scanning tunneling microscope image showing the individual atoms making up this gold (100) surface. The surface atoms deviate from the bulk crystal structure and arrange in columns several atoms wide with pits between them

The scanning tunneling microscope is a device for viewing surfaces at the atomic level. It uses the quantum tunneling phenomenon, which allows particles to pass through a barrier that would normally be insurmountable. Electrons tunnel through the vacuum between two planar metal electrodes, on each of which is an adsorbed atom, providing a tunneling-current density that can be measured. Scanning one atom (taken as the tip) as it moves past the other (the sample) permits plotting of tip displacement versus lateral separation for a constant current. The calculation shows the extent to which scanning-tunneling-microscope images of an individual atom are visible. It confirms that for low bias, the microscope images the space-averaged dimensions of the electron orbitals across closely packed energy levels—the Fermi level local density of states.

An atom can be ionized by removing one of its electrons. The electric charge causes the trajectory of an atom to bend when it passes through a magnetic field. The radius by which the trajectory of a moving ion is turned by the magnetic field is determined by the mass of the atom. The mass spectrometer uses this principle to measure the mass-to-charge ratio of ions. If a sample contains multiple isotopes, the mass spectrometer can determine the proportion of each isotope in the sample by measuring the intensity of the different beams of ions. Techniques to vaporize atoms include inductively coupled plasma atomic emission spectroscopy and inductively coupled plasma mass spectrometry, both of which use a plasma to vaporize samples for analysis.

A more area-selective method is electron energy loss spectroscopy, which measures the energy loss of an electron beam within a transmission electron microscope when it interacts with a portion of a sample. The atom-probe tomograph has sub-nanometer resolution in 3-D and can chemically identify individual atoms using time-of-flight mass spectrometry.

Spectra of excited states can be used to analyze the atomic composition of distant stars. Specific light wavelengths contained in the observed light from stars can be separated out and related to the quantized transitions in free gas atoms. These colors can be replicated using a gas-discharge lamp containing the same element. Helium was discovered in this way in the spectrum of the Sun 23 years before it was found on Earth.

Origin and Current State

Atoms form about 4% of the total energy density of the observable Universe, with an average density of about 0.25 atoms/m^3. Within a galaxy such as the Milky Way, atoms have a much higher concentration, with the density of matter in the interstellar medium (ISM) ranging from 10^5 to 10^9 atoms/m^3. The Sun is believed to be inside the Local Bubble, a region of highly ionized gas, so the density in the solar neighborhood is only about 10^3 atoms/m^3. Stars form from dense clouds in the ISM, and the evolutionary processes of stars result in the steady enrichment of the ISM with elements more massive than hydrogen and helium. Up to 95% of the Milky Way's atoms are concentrated inside stars and the total mass of atoms forms about 10% of the mass of the galaxy. (The remainder of the mass is an unknown dark matter.)

Formation

Electrons are thought to exist in the Universe since early stages of the Big Bang. Atomic nuclei forms in nucleosynthesis reactions. In about three minutes Big Bang nucleosynthesis produced most of the helium, lithium, and deuterium in the Universe, and perhaps some of the beryllium and boron.

Ubiquitousness and stability of atoms relies on their binding energy, which means that an atom has a lower energy than an unbound system of the nucleus and electrons. Where the temperature is much higher than ionization potential, the matter exists in the form of plasma—a gas of positively charged ions (possibly, bare nuclei) and electrons. When the temperature drops below the ionization potential, atoms become statistically favorable. Atoms (complete with bound electrons) became to dominate over charged particles 380,000 years after the Big Bang—an epoch called recombination, when the expanding Universe cooled enough to allow electrons to become attached to nuclei.

Since the Big Bang, which produced no carbon or heavier elements, atomic nuclei have been combined in stars through the process of nuclear fusion to produce more of the element helium, and (via the triple alpha process) the sequence of elements from carbon up to iron.

Isotopes such as lithium-6, as well as some beryllium and boron are generated in space through cosmic ray spallation. This occurs when a high-energy proton strikes an atomic nucleus, causing large numbers of nucleons to be ejected.

Elements heavier than iron were produced in supernovae through the r-process and in AGB stars through the s-process, both of which involve the capture of neutrons by atomic nuclei. Elements such as lead formed largely through the radioactive decay of heavier elements.

Earth

Most of the atoms that make up the Earth and its inhabitants were present in their current form in the nebula that collapsed out of a molecular cloud to form the Solar System. The rest are the result of radioactive decay, and their relative proportion can be used to determine the age of the Earth through radiometric dating. Most of the helium in the crust of the Earth (about 99% of the helium from gas wells, as shown by its lower abundance of helium-3) is a product of alpha decay.

There are a few trace atoms on Earth that were not present at the beginning (i.e., not "primordial"), nor are results of radioactive decay. Carbon-14 is continuously generated by cosmic rays in the atmosphere. Some atoms on Earth have been artificially generated either deliberately or as by-products of nuclear reactors or explosions. Of the transuranic elements—those with atomic numbers greater than 92—only plutonium and neptunium occur naturally on Earth. Transuranic elements have radioactive lifetimes shorter than the current age of the Earth and thus identifiable quantities of these elements have long since decayed, with the exception of traces of plutonium-244 possibly deposited by cosmic dust. Natural deposits of plutonium and neptunium are produced by neutron capture in uranium ore.

The Earth contains approximately 1.33×10^{50} atoms. Although small numbers of independent atoms of noble gases exist, such as argon, neon, and helium, 99% of the atmosphere is bound in the form of molecules, including carbon dioxide and diatomic oxygen and nitrogen. At the surface of the Earth, an overwhelming majority of atoms combine to form various compounds, including water, salt, silicates and oxides. Atoms can also combine to create materials that do not consist of discrete molecules, including crystals and liquid or solid metals. This atomic matter forms networked arrangements that lack the particular type of small-scale interrupted order associated with molecular matter.

Rare and Theoretical Forms

Superheavy Elements

While isotopes with atomic numbers higher than lead (82) are known to be radioactive, an "island of stability" has been proposed for some elements with atomic numbers above 103. These superheavy elements may have a nucleus that is relatively stable against radioactive decay. The most likely candidate for a stable superheavy atom, unbihexium, has 126 protons and 184 neutrons.

Exotic Matter

Each particle of matter has a corresponding antimatter particle with the opposite electrical charge. Thus, the positron is a positively charged antielectron and the antiproton is a negatively charged equivalent of a proton. When a matter and corresponding antimatter particle meet, they annihilate each other. Because of this, along with an imbalance between the number of matter and antimatter particles, the latter are rare in the universe. The first causes of this imbalance are not yet fully understood, although theories of baryogenesis may offer an explanation. As a result, no antimatter atoms have been discovered in nature. However, in 1996 the antimatter counterpart of the hydrogen atom (antihydrogen) was synthesized at the CERN laboratory in Geneva.

Other exotic atoms have been created by replacing one of the protons, neutrons or electrons with other particles that have the same charge. For example, an electron can be replaced by a more massive muon, forming a muonic atom. These types of atoms can be used to test the fundamental predictions of physics.

Molecule

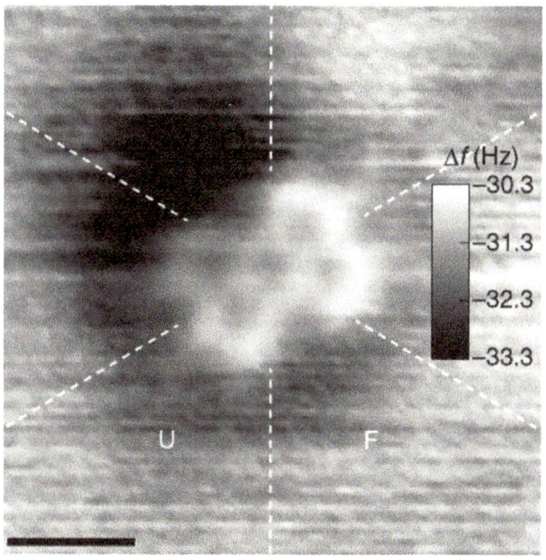

Atomic force microscopy (AFM) image of a PTCDA molecule, which contains clusters of five carbon rings

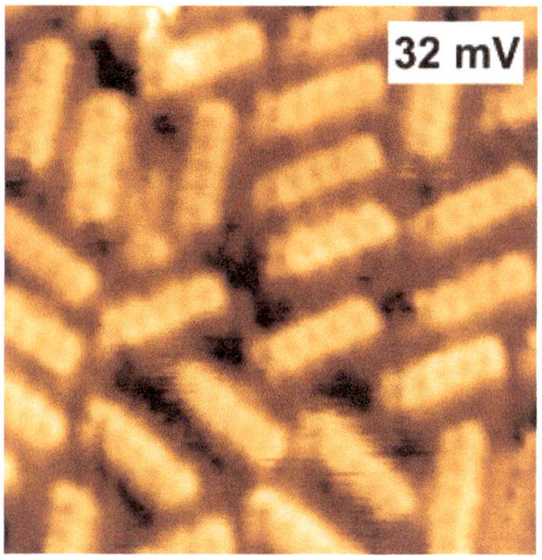

A scanning tunneling microscopy image of pentacene molecules, which consist of linear chains of five carbon rings

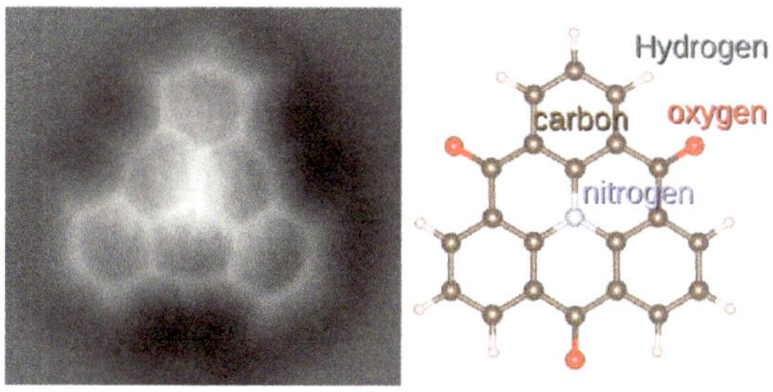

AFM image of 1,5,9-trioxo-13-azatriangulene and its chemical structure

A molecule is an electrically neutral group of two or more atoms held together by chemical bonds. Molecules are distinguished from ions by their lack of electrical charge. However, in quantum physics, organic chemistry, and biochemistry, the term *molecule* is often used less strictly, also being applied to polyatomic ions.

In the kinetic theory of gases, the term *molecule* is often used for any gaseous particle regardless of its composition. According to this definition, noble gas atoms are considered molecules as they are in fact monoatomic molecules.

A molecule may be homonuclear, that is, it consists of atoms of one chemical element, as with oxygen (O_2); or it may be heteronuclear, a chemical compound composed of more than one element, as with water (H_2O). Atoms and complexes connected by non-covalent interactions, such as hydrogen bonds or ionic bonds, are generally not considered single molecules.

Molecules as components of matter are common in organic substances (and therefore biochemistry). They also make up most of the oceans and atmosphere. However, the majority of familiar solid substances on Earth, including most of the minerals that make up the crust, mantle, and core of the Earth, contain many chemical bonds, but are *not* made of identifiable molecules. Also, no typical molecule can be defined for ionic crystals (salts) and covalent crystals (network solids), although these are often composed of repeating unit cells that extend either in a plane (such as in graphene) or three-dimensionally (such as in diamond, quartz, or sodium chloride). The theme of repeated unit-cellular-structure also holds for most condensed phases with metallic bonding, which means that solid metals are also not made of molecules. In glasses (solids that exist in a vitreous disordered state), atoms may also be held together by chemical bonds with no presence of any definable molecule, nor any of the regularity of repeating units that characterizes crystals.

Molecular Science

The science of molecules is called *molecular chemistry* or *molecular physics*, depending on whether the focus is on chemistry or physics. Molecular chemistry deals with the laws governing the interaction between molecules that results in the formation and breakage of chemical bonds, while molecular physics deals with the laws governing their structure and properties. In practice, however, this distinction is vague. In molecular sciences, a molecule consists of a stable system (bound state) composed of two or more atoms. Polyatomic ions may sometimes be usefully thought of as electrically charged molecules. The term *unstable molecule* is used for very reactive species, i.e., short-lived assemblies (resonances) of electrons and nuclei, such as radicals, molecular ions, Rydberg molecules, transition states, van der Waals complexes, or systems of colliding atoms as in Bose–Einstein condensate.

History and Etymology

According to Merriam-Webster and the Online Etymology Dictionary, the word "molecule" derives from the Latin "moles" or small unit of mass.

> Molecule (1794) – "extremely minute particle", from French molécule (1678), from New Latin molecula, diminutive of Latin *moles* "mass, barrier". A vague meaning at first; the vogue for the word (used until the late 18th century only in Latin form) can be traced to the philosophy of Descartes.

The definition of the molecule has evolved as knowledge of the structure of molecules has increased. Earlier definitions were less precise, defining molecules as the smallest particles of pure chemical substances that still retain their composition and chemical properties. This definition often breaks down since many substances in ordinary experience, such as rocks, salts, and metals, are composed of large crystalline networks of chemically bonded atoms or ions, but are not made of discrete molecules.

Bonding

Molecules are held together by either covalent bonding or ionic bonding. Several types of non-metal elements exist only as molecules in the environment. For example, hydrogen only exists as hydrogen molecule. A molecule of a compound is made out of two or more elements.

Covalent

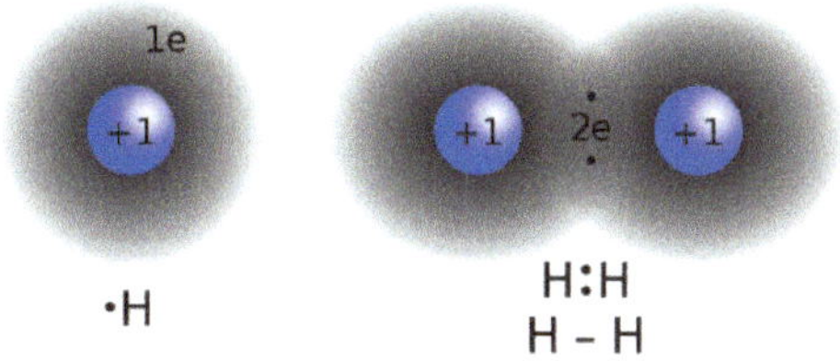

A covalent bond forming H_2 (right) where two hydrogen atoms share the two electrons

A covalent bond is a chemical bond that involves the sharing of electron pairs between atoms. These electron pairs are termed *shared pairs* or *bonding pairs*, and the stable balance of attractive and repulsive forces between atoms, when they share electrons, is termed *covalent bonding*.

Ionic

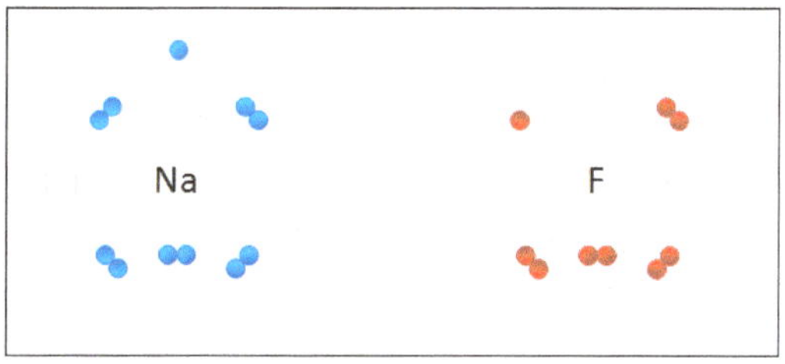

Sodium and fluorine undergoing a redox reaction to form sodium fluoride. Sodium loses its outer electron to give it a stable electron configuration, and this electron enters the fluorine atom exothermically

Ionic bonding is a type of chemical bond that involves the electrostatic attraction between oppositely charged ions, and is the primary interaction occurring in ionic compounds. The ions are atoms that have lost one or more electrons (termed cations) and atoms that have gained one or more electrons (termed anions). This transfer of electrons is termed *electrovalence* in contrast to covalence. In the simplest case, the cation is a metal atom and the anion is a nonmetal atom, but these ions can be of a more complicated nature, e.g. molecular ions like NH_4^+ or SO_4^{2-}. In simpler words, an ionic bond is the transfer of electrons from a metal to a non-metal for both atoms to obtain a full valence shell.

Molecular Size

Most molecules are far too small to be seen with the naked eye, but there are exceptions. DNA, a macromolecule, can reach macroscopic sizes, as can molecules of many polymers. Molecules commonly used as building blocks for organic synthesis have a dimension of a few angstroms (Å) to several dozen Å, or around one billionth of a meter. Single molecules cannot usually be observed by light, but small molecules and even the outlines of individual atoms may be traced in some circumstances by use of an atomic force microscope. Some of the largest molecules are macromolecules or supermolecules.

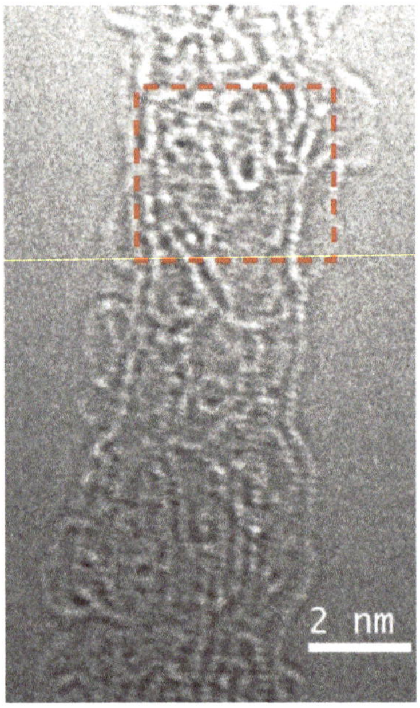

Arrangement of polyvinylidene fluoride molecules in a nanofiber – transmission electron microscopy image

The smallest molecule is the diatomic hydrogen (H_2), with a bond length of 0.74 Å.

Effective molecular radius is the size a molecule displays in solution. The table of permselectivity for different substances contains examples.

Molecular Formulas

Chemical Formula Types

The chemical formula for a molecule uses one line of chemical element symbols, numbers, and sometimes also other symbols, such as parentheses, dashes, brackets, and *plus* (+) and *minus* (–) signs. These are limited to one typographic line of symbols, which may include subscripts and superscripts.

A compound's empirical formula is a very simple type of chemical formula. It is the simplest integer ratio of the chemical elements that constitute it. For example, water is always composed of a 2:1 ratio of hydrogen to oxygen atoms, and ethyl alcohol or ethanol is always composed of carbon, hydrogen, and oxygen in a 2:6:1 ratio. However, this does not determine the kind of molecule uniquely – dimethyl ether has the same ratios as ethanol, for instance. Molecules with the same atoms in different arrangements are called isomers. Also carbohydrates, for example, have the same ratio (carbon:hydrogen:oxygen= 1:2:1) (and thus the same empirical formula) but different total numbers of atoms in the molecule.

The molecular formula reflects the exact number of atoms that compose the molecule and so characterizes different molecules. However different isomers can have the same atomic composition while being different molecules.

The empirical formula is often the same as the molecular formula but not always. For example, the molecule acetylene has molecular formula C_2H_2, but the simplest integer ratio of elements is CH.

The molecular mass can be calculated from the chemical formula and is expressed in conventional atomic mass units equal to 1/12 of the mass of a neutral carbon-12 (^{12}C isotope) atom. For network solids, the term formula unit is used in stoichiometric calculations.

Structural Formula

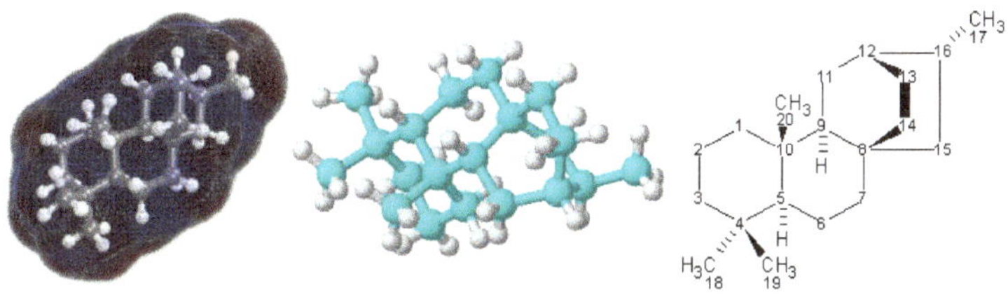

3D (left and center) and 2D (right) representations of the terpenoid molecule atisane

For molecules with a complicated 3-dimensional structure, especially involving atoms bonded to four different substituents, a simple molecular formula or even semi-structural chemical formula may not be enough to completely specify the molecule. In this case, a graphical type of formula called a structural formula may be needed. Structural formulas may in turn be represented with a one-dimensional chemical name, but such chemical nomenclature requires many words and terms which are not part of chemical formulas.

Molecular Geometry

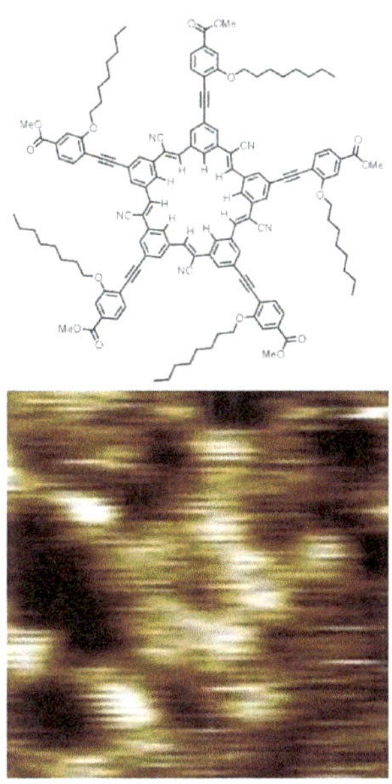

Structure and STM image of a "cyanostar" dendrimer molecule

Molecules have fixed equilibrium geometries—bond lengths and angles— about which they continuously oscillate through vibrational and rotational motions. A pure substance is composed of molecules with the same average geometrical structure. The chemical formula and the structure of a molecule are the two important factors that determine its properties, particularly its reactivity. Isomers share a chemical formula but normally have very different properties because of their different structures. Stereoisomers, a particular type of isomer, may have very similar physico-chemical properties and at the same time different biochemical activities.

Molecular Spectroscopy

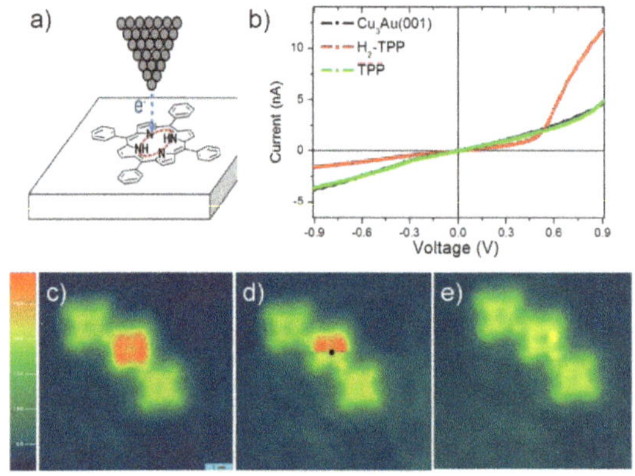

Hydrogen can be removed from individual H_2TPP molecules by applying excess voltage to the tip of a scanning tunneling microscope (STM, a); this removal alters the current-voltage (I-V) curves of TPP molecules, measured using the same STM tip, from diode like (red curve in b) to resistor like (green curve). Image (c) shows a row of TPP, H_2TPP and TPP molecules. While scanning image (d), excess voltage was applied to H_2TPP at the black dot, which instantly removed hydrogen, as shown in the bottom part of (d) and in the rescan image (e). Such manipulations can be used in single-molecule electronics

Molecular spectroscopy deals with the response (spectrum) of molecules interacting with probing signals of known energy (or frequency, according to Planck's formula). Molecules have quantized energy levels that can be analyzed by detecting the molecule's energy exchange through absorbance or emission. Spectroscopy does not generally refer to diffraction studies where particles such as neutrons, electrons, or high energy X-rays interact with a regular arrangement of molecules (as in a crystal).

Microwave spectroscopy commonly measures changes in the rotation of molecules, and can be used to identify molecules in outer space. Infrared spectroscopy measures changes in vibration of molecules, including stretching, bending or twisting motions. It is commonly used to identify the kinds of bonds or functional groups in molecules. Changes in the arrangements of electrons yield absorption or emission lines in ultraviolet, visible or near infrared light, and result in colour. Nuclear resonance spectroscopy actually measures the environment of particular nuclei in the molecule, and can be used to characterise the numbers of atoms in different positions in a molecule.

Theoretical Aspects

The study of molecules by molecular physics and theoretical chemistry is largely based on quantum mechanics and is essential for the understanding of the chemical bond. The simplest of molecules is the hydrogen molecule-ion, H_2^+, and the simplest of all the chemical bonds is the one-electron bond. H_2^+ is composed of two positively charged protons and one negatively charged electron, which means that the Schrödinger equation for the system can be solved more easily due to the lack of electron–electron repulsion. With the development of fast digital computers, approximate solutions for more complicated molecules became possible and are one of the main aspects of computational chemistry.

When trying to define rigorously whether an arrangement of atoms is *sufficiently stable* to be considered a molecule, IUPAC suggests that it "must correspond to a depression on the potential energy surface that is deep enough to confine at least one vibrational state". This definition does not depend on the nature of the interaction between the atoms, but only on the strength of the interaction. In fact, it includes weakly bound species that would not traditionally be considered molecules, such as the helium dimer, He_2, which has one vibrational bound state and is so loosely bound that it is only likely to be observed at very low temperatures.

Whether or not an arrangement of atoms is *sufficiently stable* to be considered a molecule is inherently an operational definition. Philosophically, therefore, a molecule is not a fundamental entity (in contrast, for instance, to an elementary particle); rather, the concept of a molecule is the chemist's way of making a useful statement about the strengths of atomic-scale interactions in the world that we observe.

Atomic, Molecular and Optical Physics

Atomic, molecular, and optical physics (AMO) is the study of matter-matter and light-matter interactions; at the scale of one or a few atoms and energy scales around several electron volts. The three areas are closely interrelated. AMO theory includes classical, semi-classical and quantum treatments. Typically, the theory and applications of emission, absorption, scattering of electromagnetic radiation (light) from excited atoms and molecules, analysis of spectroscopy, generation of lasers and masers, and the optical properties of matter in general, fall into these categories.

Atomic and Molecular Physics

Atomic physics is the subfield of AMO that studies atoms as an isolated system of electrons and an atomic nucleus, while molecular physics is the study of the physical properties of molecules. The term atomic physics is often associated with nuclear power and nuclear bombs, due to the synonymous use of *atomic* and *nuclear* in standard English. However, physicists distinguish between atomic physics — which deals with the atom as a system consisting of a nucleus and electrons — and nuclear physics, which considers atomic nuclei alone. The important experimental techniques are the various types of spectroscopy. Molecular physics, while closely related to atomic physics, also overlaps greatly with theoretical chemistry, physical chemistry and chemical physics.

Both subfields are primarily concerned with electronic structure and the dynamical processes by which these arrangements change. Generally this work involves using quantum mechanics. For molecular physics this approach is known as quantum chemistry. One important aspect of molecular physics is that the essential atomic orbital theory in the field of atomic physics expands to the molecular orbital theory. Molecular physics is concerned with atomic processes in molecules, but it is additionally concerned with effects due to the molecular structure. Additionally to the electronic excitation states which are known from atoms, molecules are able to rotate and to vibrate. These rotations and vibrations are quantized; there are discrete energy levels. The smallest energy differences exist between different rotational states, therefore pure rotational spectra are in the far infrared region (about 30 - 150 μm wavelength) of the electromagnetic spectrum. Vibrational spectra are in the near infrared (about 1 - 5 μm) and spectra resulting from electronic transitions are mostly in the visible and ultraviolet regions. From measuring rotational and vibrational spectra properties of molecules like the distance between the nuclei can be calculated.

As with many scientific fields, strict delineation can be highly contrived and atomic physics is often considered in the wider context of *atomic, molecular, and optical physics*. Physics research groups are usually so classified.

Optical Physics

Optical physics is the study of the generation of electromagnetic radiation, the properties of that radiation, and the interaction of that radiation with matter, especially its manipulation and control. It differs from general optics and optical engineering in that it is focused on the discovery and application of new phenomena. There is no strong distinction, however, between optical physics, applied optics, and optical engineering, since the devices of optical engineering and the applications of applied optics are necessary for basic research in optical physics, and that research leads to the development of new devices and applications. Often the same people are involved in both the basic research and the applied technology development.

Researchers in optical physics use and develop light sources that span the electromagnetic spectrum from microwaves to X-rays. The field includes the generation and detection of light, linear and nonlinear optical processes, and spectroscopy. Lasers and laser spectroscopy have transformed optical science. Major study in optical physics is also devoted to quantum optics and coherence, and to femtosecond optics. In optical physics, support is also provided in areas such as the nonlinear response of isolated atoms to intense, ultra-short electromagnetic fields, the atom-cavity interaction at high fields, and quantum properties of the electromagnetic field.

Other important areas of research include the development of novel optical techniques for nano-optical measurements, diffractive optics, low-coherence interferometry, optical coherence tomography, and near-field microscopy. Research in optical physics places an emphasis on ultrafast optical science and technology. The applications of optical physics create advancements in communications, medicine, manufacturing, and even entertainment.

History

One of the earliest steps towards *atomic physics* was the recognition that matter was composed of *atoms*, in modern terms the basic unit of a chemical element. This theory was developed by John Dalton in the 18th century. At this stage, it wasn't clear what atoms were - although they could be described and classified by their observable properties in bulk; summarized by the developing periodic table, by John Newlands and Dmitri Mendeleyev around the mid to late 19th century.

Later, the connection between atomic physics *and* optical physics became apparent, by the discovery of spectral lines and attempts to describe the phenomenon - notably by Joseph von Fraunhofer, Fresnel, and others in the 19th century.

From that time to the 1920s, physicists were seeking to explain atomic spectra and blackbody radiation. One attempt to explain hydrogen spectral lines was the Bohr atom model.

Experiments including electromagnetic radiation and matter - such as the photoelectric effect, Compton effect, and spectra of sunlight the due to the unknown element of Helium, the limitation of the Bohr model to Hydrogen, and numerous other reasons, lead to an entirely new mathematical model of matter and light: quantum mechanics.

Classical Oscillator Model of Matter

Early models to explain the origin of the index of refraction treated an electron in an atomic system classically according to the model of Paul Drude and Hendrik Lorentz. The theory was developed to attempt to provide an origin for the wavelength-dependent refractive index *n* of a material. In this model, incident electromagnetic waves forced an electron bound to an atom to oscillate. The amplitude of the oscillation would then have a relationship to the frequency of the incident electromagnetic wave and the resonant frequencies of the oscillator. The superposition of these emitted waves from many oscillators would then lead to a wave which moved more slowly.

Early Quantum Model of Matter and Light

Max Planck derived a formula to describe the electromagnetic field inside a box when in thermal equilibrium in 1900. His model consisted of a superposition of standing waves. In one dimension, the box has length *L*, and only sinusodial waves of wavenumber

$$k = \frac{n\pi}{L}$$

can occur in the box, where *n* is a positive integer (mathematically denoted by $n \in \mathbb{N}_1$). The equation describing these standing waves is given by:

$$E = E_0 \sin\left(\frac{n\pi}{L}x\right)$$

where E_0 is the magnitude of the electric field amplitude, and *E* is the magnitude of the electric field at position *x*. From this basic, Planck's law was derived.

In 1911, Ernest Rutherford concluded, based on alpha particle scattering, that an atom has a central pointlike proton. He also thought that an electron would be still attracted to the proton by Coulomb's law, which he had verified still held at small scales. As a result, he believed that electrons revolved around the proton. Niels Bohr, in 1913, combined the Rutherford model of the atom with the quantisation ideas of Planck. Only specific and well-defined orbits of the electron could exist, which also do not radiate light. In jumping orbit the electron would emit or absorb light corresponding to the difference in energy of the orbits. His prediction of the energy levels was then consistent with observation.

These results, based on a *discrete* set of specific standing waves, were inconsistent with the *continuous* classical oscillator model.

Work by Albert Einstein in 1905 on the photoelectric effect led to the association of a light wave of frequency ν with a photon of energy $h\nu$. In 1917 Einstein created an extension to Bohrs model by the introduction of the three processes of stimulated emission, spontaneous emission and absorption (electromagnetic radiation).

Modern Treatments

The largest steps towards the modern treatment was the formulation of quantum mechanics with the matrix mechanics approach by Werner Heisenberg and the discovery of the Schrödinger equation by Erwin Schrödinger.

There are a variety of semi-classical treatments within AMO. Which aspects of the problem are treated quantum mechanically and which are treated classically is dependent on the specific problem at hand. The semi-classical approach is ubiquitous in computational work within AMO, largely due to the large decrease in computational cost and complexity associated with it.

For matter under the action of a laser, a fully quantum mechanical treatment of the atomic or molecular system is combined with the system being under the action of a classical electromagnetic field. Since the field is treated classically it can not deal with spontaneous emission. This semi-classical treatment is valid for most systems, particular those under the action of high intensity laser fields. The distinction between optical physics and quantum optics is the use of semi-classical and fully quantum treatments respectively.

Within collision dynamics and using the semi-classical treatment, the internal degrees of freedom may be treated quantum mechanically, whilst the relative motion of the quantum systems under consideration are treated classically. When considering medium to high speed collisions, the nuclei can be treated classically while the electron is treated quantum mechanically. In low speed collisions the approximation fails.

Classical Monte-Carlo methods for the dynamics of electrons can be described as semi-classical in that the initial conditions are calculated using a fully quantum treatment, but all further treatment is classical.

Isolated Atoms and Molecules

Atomic, Molecular and Optical physics frequently considers atoms and molecules in isolation. Atomic models will consist of a single nucleus that may be surrounded by one or more bound electrons, whilst molecular models are typically concerned with molecular hydrogen and its molecular hydrogen ion. It is concerned with processes such as ionization, above threshold ionization and excitation by photons or collisions with atomic particles.

While modelling atoms in isolation may not seem realistic, if one considers molecules in a gas or plasma then the time-scales for molecule-molecule interactions are huge in comparison to the atomic and molecular processes that we are concerned with. This means that the individual molecules can be treated as if each were in isolation for the vast majority of the time. By this consideration atomic and molecular physics provides the underlying theory in plasma physics and atmospheric physics even though both deal with huge numbers of molecules.

Electronic Configuration

Electrons form notional shells around the nucleus. These are naturally in a ground state but can be excited by the absorption of energy from light (photons), magnetic fields, or interaction with a colliding particle (typically other electrons).

Electrons that populate a shell are said to be in a bound state. The energy necessary to remove an electron from its shell (taking it to infinity) is called the binding energy. Any quantity of energy absorbed by the electron in excess of this amount is converted to kinetic energy according to the conservation of energy. The atom is said to have undergone the process of ionization.

In the event that the electron absorbs a quantity of energy less than the binding energy, it may transition to an excited state or to a virtual state. After a statistically sufficient quantity of time, an electron in an excited state will undergo a transition to a lower state via spontaneous emission. The change in energy between the two energy levels must be accounted for (conservation of energy). In a neutral atom, the system will emit a photon of the difference in energy. However, if the lower state is in an inner shell, a phenomenon known as the Auger effect may take place where the energy is transferred to another bound electrons causing it to go into the continuum. This allows one to multiply ionize an atom with a single photon.

There are strict selection rules as to the electronic configurations that can be reached by excitation by light—however there are no such rules for excitation by collision processes.

Atomic Physics

Atomic physics is the field of physics that studies atoms as an isolated system of electrons and an atomic nucleus. It is primarily concerned with the arrangement of electrons around the nucleus and the processes by which these arrangements change. This comprises ions, neutral atoms and, unless otherwise stated, it can be assumed that the term *atom* includes ions.

The term *atomic physics* can be associated with nuclear power and nuclear weapons, due to the synonymous use of *atomic* and *nuclear* in standard English. Physicists distinguish between atomic physics — which deals with the atom as a system consisting of a nucleus and electrons — and nuclear physics, which considers atomic nuclei alone.

As with many scientific fields, strict delineation can be highly contrived and atomic physics is often considered in the wider context of *atomic, molecular, and optical physics*. Physics research groups are usually so classified.

Isolated Atoms

Atomic physics primarily considers atoms in isolation. Atomic models will consist of a single nucleus that may be surrounded by one or more bound electrons. It is not concerned with the formation of molecules (although much of the physics is identical), nor does it examine atoms in a solid state as condensed matter. It is concerned with processes such as ionization and excitation by photons or collisions with atomic particles.

While modelling atoms in isolation may not seem realistic, if one considers atoms in a gas or plasma then the time-scales for atom-atom interactions are huge in comparison to the atomic processes that are generally considered. This means that the individual atoms can be treated as if each were in isolation, as the vast majority of the time they are. By this consideration atomic physics provides the underlying theory in plasma physics and atmospheric physics, even though both deal with very large numbers of atoms.

Electronic Configuration

Electrons form notional shells around the nucleus. These are normally in a ground state but can be excited by the absorption of energy from light (photons), magnetic fields, or interaction with a colliding particle (typically ions or other electrons).

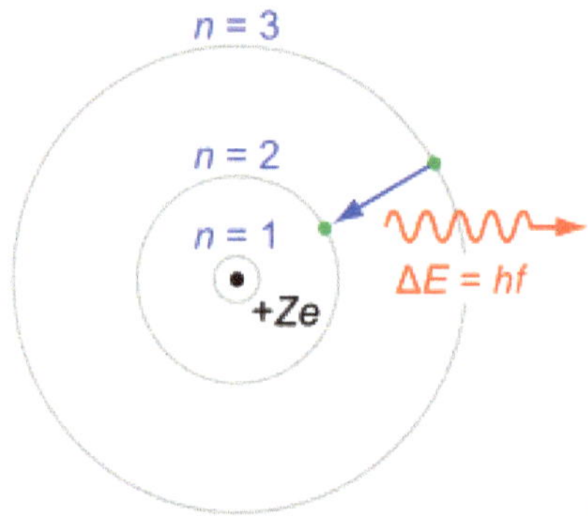

t

In the Bohr model, the transition of an electron with n=3 to the shell n=2 is shown, where a photon is emitted. An electron from shell (n=2) must have been removed beforehand by ionization

Electrons that populate a shell are said to be in a bound state. The energy necessary to remove an electron from its shell (taking it to infinity) is called the binding energy. Any quantity of energy absorbed by the electron in excess of this amount is converted to kinetic energy according to the conservation of energy. The atom is said to have undergone the process of ionization.

If the electron absorbs a quantity of energy less than the binding energy, it will be transferred to an excited state. After a certain time, the electron in an excited state will "jump" (undergo a transition) to a lower state. In a neutral atom, the system will emit a photon of the difference in energy, since energy is conserved.

If an inner electron has absorbed more than the binding energy (so that the atom ionizes), then a more outer electron may undergo a transition to fill the inner orbital. In this case, a visible photon or a characteristic x-ray is emitted, or a phenomenon known as the Auger effect may take place, where the released energy is transferred to another bound electron, causing it to go into the continuum. The Auger effect allows one to multiply ionize an atom with a single photon.

There are rather strict selection rules as to the electronic configurations that can be reached by excitation by light — however there are no such rules for excitation by collision processes.

History and Developments

The majority of fields in physics can be divided between theoretical work and experimental work, and atomic physics is no exception. It is usually the case, but not always, that progress goes in alternate cycles from an experimental observation, through to a theoretical explanation followed by some predictions that may or may not be confirmed by experiment, and so on. Of course, the current state of technology at any given time can put limitations on what can be achieved experimentally and theoretically so it may take considerable time for theory to be refined.

One of the earliest steps towards atomic physics was the recognition that matter was composed of *atoms*. It forms a part of the texts written in 6th century BC to 2nd century BC such as those of Democritus or Vaisheshika Sutra written by Kanad. This theory was later developed in the modern sense of the basic unit of a chemical element by the British chemist and physicist John Dalton in the 18th century. At this stage, it wasn't clear what atoms were although they could be described and classified by their properties (in bulk). The invention of the periodic system of elements by Mendeleev was another great step forward.

The true beginning of atomic physics is marked by the discovery of spectral lines and attempts to describe the phenomenon, most notably by Joseph von Fraunhofer. The study of these lines led to the Bohr atom model and to the birth of quantum mechanics. In seeking to explain atomic spectra an entirely new mathematical model of matter was revealed. As far as atoms and their electron shells were concerned, not only did this yield a better overall description, i.e. the atomic orbital model, but it also provided a new theoretical basis for chemistry (quantum chemistry) and spectroscopy.

Since the Second World War, both theoretical and experimental fields have advanced at a rapid pace. This can be attributed to progress in computing technology, which has allowed larger and more sophisticated models of atomic structure and associated collision processes. Similar technological advances in accelerators, detectors, magnetic field generation and lasers have greatly assisted experimental work.

Molecular Physics

Molecular physics is the study of the physical properties of molecules, the chemical bonds between atoms as well as the molecular dynamics. Its most important experimental techniques are the various types of spectroscopy; scattering is also used. The field is closely related to atomic physics and overlaps greatly with theoretical chemistry, physical chemistry and chemical physics.

In addition to the electronic excitation states which are known from atoms, molecules exhibit rotational and vibrational modes whose energy levels are quantized. The smallest energy differences exist between different rotational states: pure rotational spectra are in the far infrared region (about 30 - 150 μm wavelength) of the electromagnetic spectrum. Vibrational spectra are in the near infrared (about 1 - 5 μm) and spectra resulting from electronic transitions are mostly in the visible and ultraviolet regions. From measuring rotational and vibrational spectra properties of molecules like the distance between the nuclei can be calculated.

One important aspect of molecular physics is that the essential atomic orbital theory in the field of atomic physics expands to the molecular orbital theory.

Various scales in AM Physics

(Temperature range from 10^4 K to 10^{-9} K)

$\sim 10^4$ K

Fusion Plasma and stellar environment, Spectroscopy and dynamics of highly charged ions, Diagnostic techniques for plasma

~ 10K

Molecular interactions in astrophysical ices, Formation of complex molecules in interstellar environment, Fundamental question on formation of molecules which have relevance to origin of life

$\sim 10^{-1}$K

Atomic and Molecular beams, Clusters

$\sim 10^{-6}$K

Laser cooled atoms, Atomic clock based on Cs fountain, Nano-fabrication using cold atoms,

$\sim 10^{-9}$K

Quantum degenerate atomic gases, Bose-Einstein Condensation, Fermi Sea, Quantum many body physics, Matter wave optics, Atom interferometry, Atomic Physics as basis of condensed matter physics

Photon Intensities used for Interaction:

$10^{-9}–10^{-3}(\mathrm{W/cm^2}) \Rightarrow$ Obtained from normal light sources

High resolution spectroscopy for structure of atoms and molecules

$10^{3}-10^{9}(\mathrm{W/cm^2}) \Rightarrow$ Obtained from laser sources

Multi-photon processes, Multi-step excitation processes, Nonlinear optical processes, Rydberg and Autoionizing Levels, Ultra-trace analysis (single atom detection),

$10^{9}-10^{15}(\mathrm{W/cm^2}) \Rightarrow$ High intense laser pulse sources

Intense field regime, Above threshold ionization, Multiple ionization, High harmonic generation (up to 300), Coulomb explosion, Atomic Stabilization

Spectroscopy

Analysis of white light by dispersing it with a prism is an example of spectroscopy

Spectroscopy is the study of the interaction between matter and electromagnetic radiation. Historically, spectroscopy originated through the study of visible light dispersed according to its wavelength, by a prism. Later the concept was expanded greatly to include any interaction with radiative energy as a function of its wavelength or frequency. Spectroscopic data is often represented by an emission spectrum, a plot of the response of interest as a function of wavelength or frequency.

Introduction

Spectroscopy and spectrography are terms used to refer to the measurement of radiation intensity as a function of wavelength and are often used to describe experimental spectroscopic methods. Spectral measurement devices are referred to as spectrometers, spectrophotometers, spectrographs or spectral analyzers.

Daily observations of color can be related to spectroscopy. Neon lighting is a direct application of atomic spectroscopy. Neon and other noble gases have characteristic emission frequencies (colors). Neon lamps use collision of electrons with the gas to excite these emissions. Inks, dyes and paints include chemical compounds selected for their spectral characteristics in order to generate specific colors and hues. A commonly encountered molecular spectrum is that of nitrogen dioxide. Gaseous nitrogen dioxide has a characteristic red absorption feature, and this gives air polluted with nitrogen dioxide a reddish-brown color. Rayleigh scattering is a spectroscopic scattering phenomenon that accounts for the color of the sky.

Spectroscopic studies were central to the development of quantum mechanics and included Max Planck's explanation of blackbody radiation, Albert Einstein's explanation of the photoelectric effect and Niels Bohr's explanation of atomic structure and spectra. Spectroscopy is used in physical and analytical chemistry because atoms and molecules have unique spectra. As a result, these spectra can be used to detect, identify and quantify information about the atoms and molecules. Spectroscopy is also used in astronomy and remote sensing on earth. Most research telescopes have spectrographs. The measured spectra are used to determine the chemical composition and physical properties of astronomical objects (such as their temperature and velocity).

Theory

One of the central concepts in spectroscopy is a resonance and its corresponding resonant frequency. Resonances were first characterized in mechanical systems such as pendulums. Mechanical systems that vibrate or oscillate will experience large amplitude oscillations when they are driven at their resonant frequency. A plot of amplitude vs. excitation frequency will have a peak centered at the resonance frequency. This plot is one type of spectrum, with the peak often referred to as a spectral line, and most spectral lines have a similar appearance.

In quantum mechanical systems, the analogous resonance is a coupling of two quantum mechanical stationary states of one system, such as an atom, via an oscillatory source of energy such as a photon. The coupling of the two states is strongest when the energy of the source matches the energy difference between the two states. The energy (E) of a photon is related to its frequency (ν) by $E = h\nu$ where h dis Planck's constant, and so a spectrum of the system response vs. photon frequency will peak at the resonant frequency or energy. Particles such as electrons and neutrons have a comparable relationship, the de Broglie relations, between their kinetic energy and their wavelength and frequency and therefore can also excite resonant interactions.

Spectra of atoms and molecules often consist of a series of spectral lines, each one representing a resonance between two different quantum states. The explanation of these series, and the spectral patterns associated with them, were one of the experimental enigmas that drove the development and acceptance of quantum mechanics. The hydrogen spectral series in particular was first successfully explained by the Rutherford-Bohr quantum model of the hydrogen atom. In some cases spectral lines are well separated and distinguishable, but spectral lines can also overlap and appear to be a single transition if the density of energy states is high enough. Named series of lines include the principal, sharp, diffuse and fundamental series.

Classification of Methods

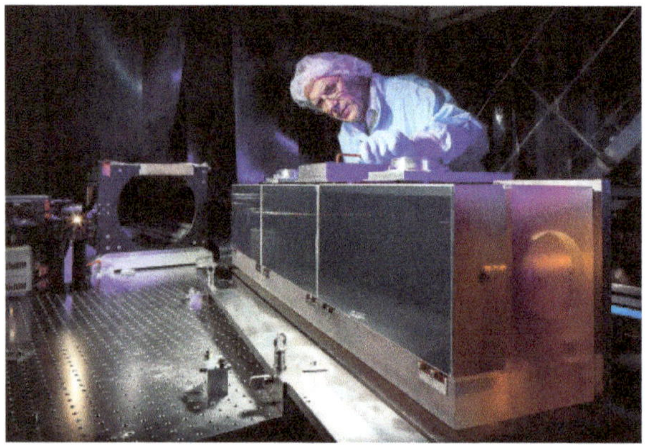

A huge diffraction grating at the heart of the ultra-precise ESPRESSO spectrograph

Spectroscopy is a sufficiently broad field that many sub-disciplines exist, each with numerous implementations of specific spectroscopic techniques. The various implementations and techniques can be classified in several ways.

Type of Radiative Energy

Types of spectroscopy are distinguished by the type of radiative energy involved in the interaction. In many applications, the spectrum is determined by measuring changes in the intensity or frequency of this energy. The types of radiative energy studied include:

- Electromagnetic radiation was the first source of energy used for spectroscopic studies. Techniques that employ electromagnetic radiation are typically classified by the wavelength region of the spectrum and include microwave, terahertz, infrared, near infrared, visible and ultraviolet, x-ray and gamma spectroscopy.
- Particles, due to their de Broglie wavelength, can also be a source of radiative energy and both electrons and neutrons are commonly used. For a particle, its kinetic energy determines its wavelength.
- Acoustic spectroscopy involves radiated pressure waves.
- Mechanical methods can be employed to impart radiating energy, similar to acoustic waves, to solid materials.

Nature of the Interaction

Types of spectroscopy can also be distinguished by the nature of the interaction between the energy and the material. These interactions include:

- Absorption occurs when energy from the radiative source is absorbed by the material. Absorption is often determined by measuring the fraction of energy transmitted through the material; absorption will decrease the transmitted portion.
- Emission indicates that radiative energy is released by the material. A material's blackbody spectrum is a spontaneous emission spectrum determined by its temperature; this feature can be measured in the infrared by instruments such as the Atmospheric Emitted Radiance Interferometer (AERI). Emission can also be induced by other sources of energy such as flames or sparks or electromagnetic radiation in the case of fluorescence.
- Elastic scattering and reflection spectroscopy determine how incident radiation is reflected or scattered by a material. Crystallography employs the scattering of high energy radiation, such as x-rays and electrons, to examine the arrangement of atoms in proteins and solid crystals.
- Impedance spectroscopy studies the ability of a medium to impede or slow the transmittance of energy. For optical applications, this is characterized by the index of refraction.
- Inelastic scattering phenomena involve an exchange of energy between the radiation and the matter that shifts the wavelength of the scattered radiation. These include Raman and Compton scattering.
- Coherent or resonance spectroscopy are techniques where the radiative energy couples two quantum states of the material in a coherent interaction that is sustained by the radiating field. The coherence can be disrupted by other interactions, such as particle collisions and energy transfer, and so often require high intensity radiation to be sustained. Nuclear magnetic resonance (NMR) spectroscopy is a widely used resonance method and ultrafast laser methods are also now possible in the infrared and visible spectral regions.

Type of Material

Spectroscopic studies are designed so that the radiant energy interacts with specific types of matter.

Atoms

Atomic spectroscopy was the first application of spectroscopy developed. Atomic absorption spectroscopy (AAS) and atomic emission spectroscopy (AES) involve visible and ultraviolet light. These absorptions and emissions, often referred to as atomic spectral lines, are due to electronic transitions of outer shell electrons as they rise and fall from one electron orbit to another. Atoms also have distinct x-ray spectra that are attributable to the excitation of inner shell electrons to excited states.

Atoms of different elements have distinct spectra and therefore atomic spectroscopy allows for the identification and quantitation of a sample's elemental composition. Robert Bunsen and Gustav Kirchhoff discovered new elements by observing their emission spectra. Atomic absorption lines are observed in the solar spectrum and referred to as Fraunhofer lines after their discoverer. A comprehensive explanation of the hydrogen spectrum was an early success of quantum mechanics and explained the Lamb shift observed in the hydrogen spectrum, which further led to the development of quantum electrodynamics.

Modern implementations of atomic spectroscopy for studying visible and ultraviolet transitions include flame emission spectroscopy, inductively coupled plasma atomic emission spectroscopy, glow discharge spectroscopy, microwave induced plasma spectroscopy, and spark or arc emission spectroscopy. Techniques for studying x-ray spectra include X-ray spectroscopy and X-ray fluorescence (XRF).

Molecules

The combination of atoms into molecules leads to the creation of unique types of energetic states and therefore unique spectra of the transitions between these states. Molecular spectra can be obtained due to electron spin states (electron paramagnetic resonance), molecular rotations, molecular vibration and electronic states. Rotations are collective motions of the atomic nuclei and typically lead to spectra in the microwave and millimeter-wave spectral regions; rotational spectroscopy and microwave spectroscopy are synonymous. Vibrations are relative motions of the atomic nuclei and are studied by both infrared and Raman spectroscopy. Electronic excitations are studied using visible and ultraviolet spectroscopy as well as fluorescence spectroscopy.

Studies in molecular spectroscopy led to the development of the first maser and contributed to the subsequent development of the laser.

Crystals and Extended Materials

The combination of atoms or molecules into crystals or other extended forms leads to the creation of additional energetic states. These states are numerous and therefore have a high density of states. This high density often makes the spectra weaker and less distinct, i.e., broader. For instance, blackbody radiation is due to the thermal motions of atoms and molecules within a material. Acoustic and mechanical responses are due to collective motions as well. Pure crystals, though, can have distinct spectral transitions, and the crystal arrangement also has an effect on the observed molecular spectra. The regular lattice structure of crystals also scatters x-rays, electrons or neutrons allowing for crystallographic studies.

Nuclei

Nuclei also have distinct energy states that are widely separated and lead to gamma ray spectra. Distinct nuclear spin states can have their energy separated by a magnetic field, and this allows for NMR spectroscopy.

Other Types

Other types of spectroscopy are distinguished by specific applications or implementations:

- Acoustic resonance spectroscopy is based on sound waves primarily in the audible and ultrasonic regions

- Auger spectroscopy is a method used to study surfaces of materials on a micro-scale. It is often used in connection with electron microscopy.
- Cavity ring down spectroscopy
- Circular Dichroism spectroscopy
- Coherent anti-Stokes Raman spectroscopy (CARS) is a recent technique that has high sensitivity and powerful applications for *in vivo* spectroscopy and imaging.
- Cold vapour atomic fluorescence spectroscopy
- Correlation spectroscopy encompasses several types of two-dimensional NMR spectroscopy.
- Deep-level transient spectroscopy measures concentration and analyzes parameters of electrically active defects in semiconducting materials
- Dual polarisation interferometry measures the real and imaginary components of the complex refractive index
- Electron phenomenological spectroscopy measures physicochemical properties and characteristics of electronic structure of multicomponent and complex molecular systems.
- EPR spectroscopy
- Force spectroscopy
- Fourier transform spectroscopy is an efficient method for processing spectra data obtained using interferometers. Fourier transform infrared spectroscopy (FTIR) is a common implementation of infrared spectroscopy. NMR also employs Fourier transforms.
- Hadron spectroscopy studies the energy/mass spectrum of hadrons according to spin, parity, and other particle properties. Baryon spectroscopy and meson spectroscopy are both types of hadron spectroscopy.
- Hyperspectral imaging is a method to create a complete picture of the environment or various objects, each pixel containing a full visible, VNIR, NIR, or infrared spectrum.
- Inelastic electron tunneling spectroscopy (IETS) uses the changes in current due to inelastic electron-vibration interaction at specific energies that can also measure optically forbidden transitions.
- Inelastic neutron scattering is similar to Raman spectroscopy, but uses neutrons instead of photons.
- Laser-Induced Breakdown Spectroscopy (LIBS), also called Laser-induced plasma spectrometry (LIPS)
- Laser spectroscopy uses tunable lasers and other types of coherent emission sources, such as optical parametric oscillators, for selective excitation of atomic or molecular species.

- Mass spectroscopy is an historical term used to refer to mass spectrometry. Current recommendations are to use the latter term. Use of the term mass spectroscopy originated in the use of phosphor screens to detect ions.
- Mössbauer spectroscopy probes the properties of specific isotopic nuclei in different atomic environments by analyzing the resonant absorption of gamma-rays.
- Neutron spin echo spectroscopy measures internal dynamics in proteins and other soft matter systems
- Photoacoustic spectroscopy measures the sound waves produced upon the absorption of radiation.
- Photoemission spectroscopy
- Photothermal spectroscopy measures heat evolved upon absorption of radiation.
- Pump-probe spectroscopy can use ultrafast laser pulses to measure reaction intermediates in the femtosecond timescale.
- Raman optical activity spectroscopy exploits Raman scattering and optical activity effects to reveal detailed information on chiral centers in molecules.
- Raman spectroscopy
- Saturated spectroscopy
- Scanning tunneling spectroscopy
- Spectrophotometry
- Time-resolved spectroscopy measures the decay rate(s) of excited states using various spectroscopic methods.
- Time-Stretch Spectroscopy
- Thermal infrared spectroscopy measures thermal radiation emitted from materials and surfaces and is used to determine the type of bonds present in a sample as well as their lattice environment. The techniques are widely used by organic chemists, mineralogists, and planetary scientists.
- Transient grating spectroscopy measures quasiparticle propagation. It can track changes in metallic materials as they are irradiated.
- Ultraviolet photoelectron spectroscopy (UPS)
- Ultraviolet–visible spectroscopy
- Vibrational circular dichroism spectroscopy
- Video spectroscopy
- X-ray photoelectron spectroscopy (XPS)

Applications

UVES is a high-resolution spectrograph on the Very Large Telescope

- Cure monitoring of composites using optical fibers.
- Estimate weathered wood exposure times using near infrared spectroscopy.
- Measurement of different compounds in food samples by absorption spectroscopy both in visible and infrared spectrum.

History

The history of spectroscopy began with Isaac Newton's optics experiments (1666–1672). Newton applied the word "spectrum" to describe the rainbow of colors that combine to form white light and that are revealed when the white light is passed through a prism. During the early 1800s, Joseph von Fraunhofer made experimental advances with dispersive spectrometers that enabled spectroscopy to become a more precise and quantitative scientific technique. Since then, spectroscopy has played and continues to play a significant role in chemistry, physics and astronomy.

- "In 1672, in the first paper that he submitted to the Royal Society, Sir Isaac Newton described an experiment in which he permitted sunlight to pass through a small hole and then through a prism. Newton found that sunlight, which looks white to us, is actually made up of a mixture of all the colors of the rainbow"
- "In 1802, William Wollaston built an improved spectrometer that included a lens to focus the Sun's spectrum on a screen. Upon use, Wollaston realized that the colors were not spread uniformly, but instead had missing patches of colors, which appeared as dark bands in the spectrum. Later, in 1815, German physicist Joseph Fraunhofer also examined the solar spectrum, and found about 600 such dark lines (missing colors), which are now known as Fraunhofer lines, or Absorption lines."

Spectral Resolution used for Spectroscopy

10 Hz

Ultra precision measurements, Atomic Clocks of accuracy 1 in 10^{16} and better, Ultra-precision magnetometry (10^{-15} Tesla detection)

$\sim 100\text{Hz} - 10^6\,\text{Hz}$

Parity violation experiments in atoms and molecules (non-accelerator particle physics), Study of hyperfine interactions, Hyperfine structure and Isotope shifts, Measurement of properties of unstable nuclei (Nuclear Physics via Atomic physics), Cold atom physics

$\sim 10^6 - 10^9\,\text{Hz}$

Fine structure in atoms and molecules, Atomic & molecular Interactions, State selective processes in atoms and molecules

$\sim 10^9 - 10^{15}\,\text{Hz}$

Spectrochemical analysis, Material characterization, Photoionization and photo-fragmentation, supercontinuum sources and their interactions

Excitation Sources used for Modern Research:

Table-top instruments to high energy particle accelerator

Lasers: 10^{-3} eV to 6 eV photons, used for Rotational, Vibrational and Electronic Spectroscopy

Synchrotron Radiation Sources: 5 eV to 1000 eV photons: used to study Valence to Inner-shell processes

Ion Traps and Atom Trap: Used to study precision measurement in neutral atoms to highly charged ions, U+91 in electron beam ion trap, Study of non-neutral plasma

Particle Accelerators: This is used to study the interactions of charged particles of MeV energies with atoms and molecules

Systems used for Investigations

A few atoms and ions : Precision measurements on unstable nuclei produced in an accelerator, Quantum Information Processing using atoms

Small sized clusters : Metallic and semiconductor clusters, $Au_{40}, Si_{20}Ge_{20,}$ Approach to bulk properties, Basic for functional materials

Large sized clusters : $(H_2o)_n, (Xe)_n$, etc, n = 1000 - 5000

Used to understand the Van der Waals complexes, solvent interactions in gas phase

10^3 cold atoms or ions : Cold atom physics, Bose Einstein Condensation etc

Gas/liquid phase : Spectroscopic analysis, trace analysis, pollution detection, chemical and biochemical interactions

In the following, some revolutions in the field of research and applications are highlighted:

- The advent of laser opens up a new arena in many disciplines of science and technology.

- The power and precision of lasers have revolutionized the field of atomic physics and molecular physics.

For example, lasers have made possible entirely new technologies for isolating individual atoms in electromagnetic traps and cooling them to near absolute zero. This is known as Laser Atom Cooling. The Nobel Prize in Physics 1997 was awarded jointly to Steven Chu, Claude Cohen-Tannoudji and William D. Phillips *"for development of methods to cool and trap atoms with laser light"*.

Briefly, when the atoms are brought essentially to rest in the trap by using laser, they can undergo a quantum mechanical phase transition to form a superfluid.

This is known as a Bose-Einstein condensation, while remaining in the form of a dilute gas. This is a new state of matter where all the atoms are in the same coherent quantum state. As a consequence, the atoms lose their individual identities, and their quantum mechanical wavelike properties become dominant.

This coherent beam of atoms can be used to form an "atom laser" analogous to the coherent beam of photons in a conventional laser.

The potential of this future technology hopefully can be exploited for the fabrication of microelectronic and other nanoscale devices.

This can well be used for the development of quantum computer, the dream of many scientists.

Due to the high precision of lasers, it dramatically increased the precision with which the characteristic wavelengths of atoms can be measured.

Modern standards of time and frequency are based on measurements of transition frequencies in atomic cesium.

An atomic clock is an oscillator whose frequency is locked to the frequency of an atomic transition.

The second is the duration of 9 192 631 770 periods of the radiation corresponding to the transition between the two hyperfine sublevels F = 4 and F = 3 of the ground electronic state of cesium 133.

As shown in figure below , the two sublevels correspond to parallel and anti parallel orientations of the spins S = 1/2 of the valence electron and I = 7/2 of the nucleus of the Cs atom.

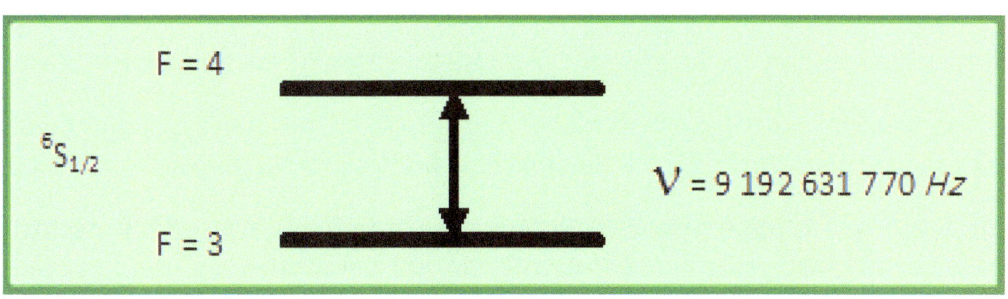

In 1930s, I.I. Rabi and his students, at Columbia University invented this technique known as magnetic resonance, by which they could measure the natural resonant frequencies of atoms. Rabi won the 1944 Nobel Prize for his work.

In 1949, Rabi's student Norman Ramsey suggested that making the atoms pass twice through the oscillating electromagnetic field could result in a much more accurate clock. In 1989 Ramsey was awarded the Nobel Prize for his work.

The discovery and the technology development of this atomic clock lead to the most important application in today's world, a global system of navigation known as Global Positioning System (GPS).

Researchers at the Lincoln Laboratory of the Massachusetts Institute of Technology (MIT) were able to determine the satellite's orbit precisely by observing Doppler shift of its radio signal. Since the satellite's orbit could be precisely determined from the ground, it established that positions on the ground could also be determined by homing in on the signals broadcast by satellites.

On June 26, 1993, U.S. Air Force launched the 24th Navstar satellite into orbit, completing a network of 24 satellites known as the Global Positioning System, or GPS.

This incredible new technology was made possible by using the world's most accurate timepieces: atomic clocks that are precise to within a billionth of a second.

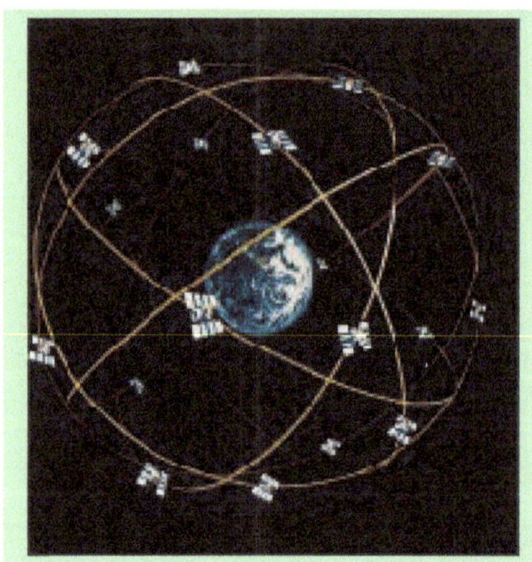

In this system as shown in the above figure , there are 24 satellites orbiting 11,000 miles above Earth emits coded signals. Four atomic clocks in each satellite keep accurate time.

Using Doppler shift, receiver on the earth's surface calculates latitude, longitude, altitude, and time by comparing signals from satellites.

This determines the location on the ground, accurate to within 30 meters, or 100 feet.

The Atomic spectroscopy is an important tool in Astrophysics. The analysis of the chemical composition from the spectrum recorded from the heavenly bodies proved many theoretical predictions.

As shown in figure below, Supernova SN1987A was seen in February 1987. It occurred in Large Magellanic Cloud, 163,000 years ago. Observing the star before the explosion proved the theory that core-collapse of a massive star produces supernova. The spectrum observed from this supernova showed the composition of the star's outer layer. The analysis revealed an overabundance of nitrogen relative to oxygen and carbon as compared to the sun which is the signature of CNO hydrogen fusion cycle.

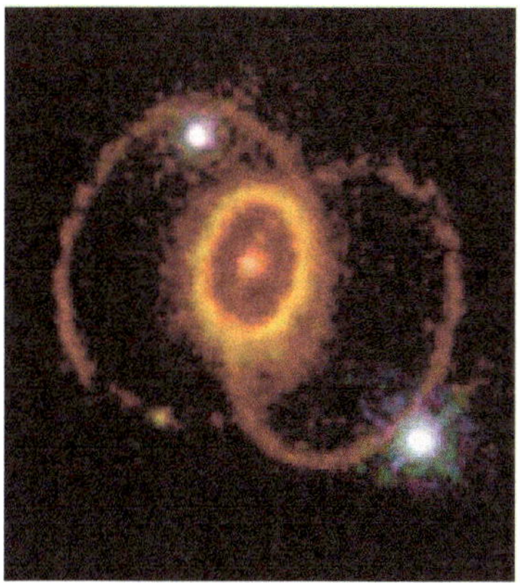

Plasma, the fourth state of matter is an interesting research field to the scientist. Creating super hot solid plasma state is a challenging task and provides greater understanding of fusion processes.

Researchers working at the U.S. Department of Energy's (DOE) SLAC National Accelerator Laboratory have used the world's most powerful X-ray laser to build the first atomic X-ray laser pulse, as well as to create in a controlled way the superheat (2-million-degree) piece of matter.

The atomic laser could be used to understand the working of biological molecules.

The creation of hot dense matter by intense laser pulse could be exploited to understand the processes of nuclear fusion that power the sun.

Molecular spectrum is the signature and the fingerprint to identify the molecules. This has a deeper and direct application to our human society.

Now-a-days several spectroscopic methods are used in medical science for diagnosis the disease. In the following, a glimpse of these methodologies is given.

X-rays: the most common instrument used almost everywhere.

The first X-ray device was discovered accidentally by the German scientist Wilhelm Roentgen (1845-1923) in 1895.

He found that invisible rays emitted from cathode-ray tube that penetrate paper and wood but could not penetrate metal and bone. Roentgen used his device to examine the bone structure of the human hand.

X-rays are produced when a target metal (such as tungsten) is bombarded by highly accelerated electrons. These high energy electrons knock out the electrons from the deep energy levels such as K and L shells, the other electrons in higher energy levels occupy these vacant spaces by radiating extra energy in the form of X-rays.

X-rays penetrate the body and the image is stored on silver halide films.

Bones and metals absorb or reflect X-rays – leaving the impression on the film, whereas the soft tissues transmit X-rays. This contrast produces an image of bone or metal deep inside the body.

This helps to identify medical problems inside the body such as fractures of bones etc.

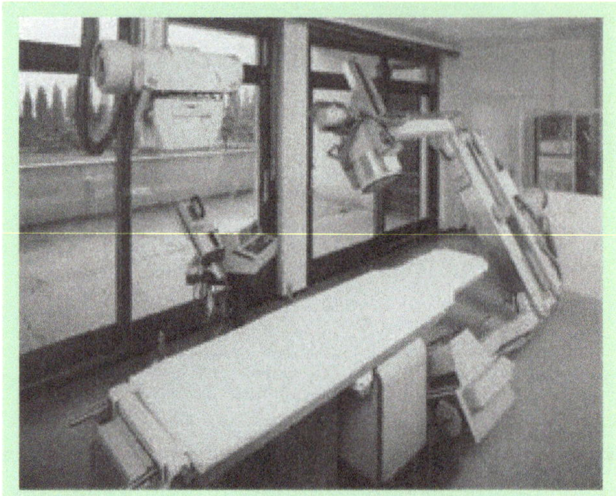

Atomic Spectroscopy

Atomic spectroscopy is the study of the electromagnetic radiation absorbed and emitted by atoms. Since unique elements have characteristic (signature) spectra, atomic spectroscopy, specifically the electromagnetic spectrum or mass spectrum, is applied for determination of elemental compositions. It can be divided by atomization source or by the type of spectroscopy used. In the latter case, the main division is between optical and mass spectrometry. Mass spectrometry generally gives significantly better analytical performance, but is also significantly more complex. This complexity translates into higher purchase costs, higher operational costs, more operator training, and a greater number of components that can potentially fail. Because optical spectroscopy is often less expensive and has performance adequate for many tasks, it is far more common Atomic absorption spectrometers are one of the most commonly sold and used analytical devices.

Optical Spectroscopy

Electrons exist in energy levels (i.e. atomic orbitals) within an atom. Atomic orbitals are quantized, meaning they exist as defined values instead of being continuous. Electrons may move between orbitals, but in doing so they must absorb or emit energy equal to the energy difference between their atom's specific quantized orbital energy levels. In optical spectroscopy, energy absorbed to move an electron to a higher energy level (higher orbital) and/or the energy emitted as the electron moves to a lower energy level is absorbed or emitted in the form of photons (light particles). Because each element has a unique number of electrons, an atom will absorb/release energy in a pattern unique to its elemental identity (e.g. Ca, Na, etc.) and thus will absorb/emit photons in a correspondingly unique pattern. The type of atoms present in a sample, or the amount of atoms present in a sample can be deduced from measuring these changes in light wavelength and light intensity.

Optical spectroscopy is further divided into atomic absorption spectroscopy and atomic emission spectroscopy. In atomic absorption spectroscopy, light of a predetermined wavelength is passed through a collection of atoms. If the wavelength of the source light has energy corresponding to the energy difference between two energy levels in the atoms, a portion of the light will be absorbed. The difference between the intensity of the light emitted from the source (e.g., lamp) and the light collected by the detector yields an absorbance value. This absorbance value can then be used to determine the concentration of a given element (or atoms) within the sample. The relationship between the concentration of atoms, the distance the light travels through the collection of atoms, and the portion of the light absorbed is given by the Beer–Lambert law. In atomic emission spectroscopy, the intensity of the emitted light is directly proportional to the concentration of atoms.

Mass Spectrometry

Atomic mass spectrometry is similar to other types of mass spectrometry in that it consists of an ion source, a mass analyzer, and a detector. Atoms' identities are determined by their mass-to-charge ratio (via the mass analyzer) and their concentrations are determined by the number of ions detected. Although considerable research has gone into customizing mass spectrometers for atomic ion sources, it is the ion source that differs most from other forms of mass spectrometry. These ion sources must also atomize samples, or an atomization step must take place before ionization. Atomic ion sources are generally modifications of atomic optical spectroscopy atom sources.

Ion and Atom Sources

Sources can be adapted in many ways. Of these, flames are the most common due to their low cost and their simplicity. Although significantly less common, inductively-coupled plasmas, especially when used with mass spectrometers, are recognized for their outstanding analytical performance and their versatility.

For all atomic spectroscopy, a sample must be vaporized and atomized. For atomic mass spectrometry, a sample must also be ionized. Vaporization, atomization, and ionization are often, but not always, accomplished with a single source. Alternatively, one source may be used to vaporize a sample while another is used to atomize (and possibly ionize). An example of this is laser ablation inductively-coupled plasma atomic emission spectrometry, where a laser is used to vaporize a solid sample and an inductively-coupled plasma is used to atomize the vapor.

With the exception of flames and graphite furnaces, which are most commonly used for atomic absorption spectroscopy, most sources are used for atomic emission spectroscopy.

Liquid-sampling sources include flames and sparks (atom source), inductively-coupled plasma (atom and ion source), graphite furnace (atom source), microwave plasma (atom and ion source), and direct-current plasma (atom and ion source). Solid-sampling sources include lasers (atom and vapor source), glow discharge (atom and ion source), arc (atom and ion source), spark (atom and ion source), and graphite furnace (atom and vapor source). Gas-sampling sources include flame (atom source), inductively-coupled plasma (atom and ion source), microwave plasma (atom and ion source), direct-current plasma (atom and ion source), and glow discharge (atom and ion source).

CT Scan or CAT Scan

The scanning mechanism is incorporated by involving a moving target or by revolving the X-ray source. The scintillator, detector for X-ray, is used to pick up the signal and produced the image. By superimposing several images, 2-dimensional and 3-dimensional images are computed.

Positron Emission Tomography

This device measures the metabolic activity by measuring the decay of the radioactive traces injected from outside in tissues. Fluorine 18 is used as radioactive element.

Diseased organs process the injected FluroDeoxyGlucose (FDG) at a higher rate than the normal tissues. Positrons are emitted by FDG and emit g-radiation when collide with electron. g-detectors pick up the signal and compute a 3-dimensional image of the body.

In this way, this methodology helps to identify the location of the diseased organs deep inside the body

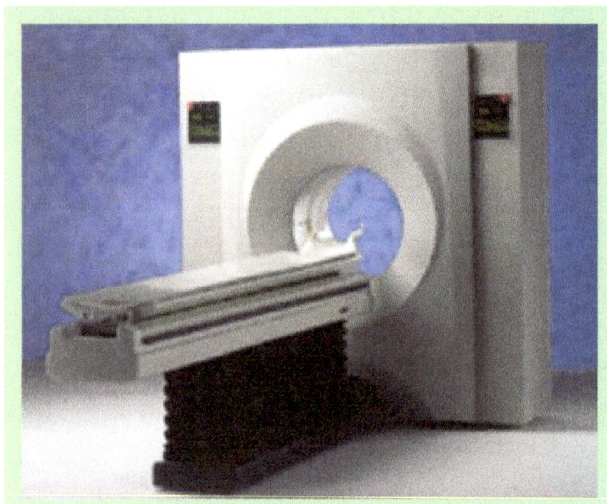

As shown in the above figure, the scanner is typically a large, box-like machine with a hole, or short tunnel, in the center. The X-ray tube is rotated and electronic detectors are located opposite each other in a ring, called a gantry. The computer workstation processes the imaging information.

Raman Spectroscopy

When the monochromatic light interacts with molecular vibrations, phonons or other excitations in the system, it results in the energy of the photons being shifted up or down. The shift in energy gives information about the vibrational modes in the system.

The Raman Effect occurs when light interacts with a molecule. The light excites the molecule from the ground state to a virtual energy state. When the molecule relaxes it emits a photon and it returns to a different rotational or vibrational state. The difference in energy between the original state and this new state leads to a shift in the emitted photon's frequency away from the excitation wavelength.

Raman spectroscopy is an invaluable technique for investigations of pharmaceutically and medically relevant molecules. It is used in many fields where non-destructive microscopic chemical analysis and imaging is required.

Both Biologists and Chemists use Raman Spectroscopy to identify chemical compounds, their functional groups, and to determine the conformation of complex biomolecules, such as proteins and DNA.

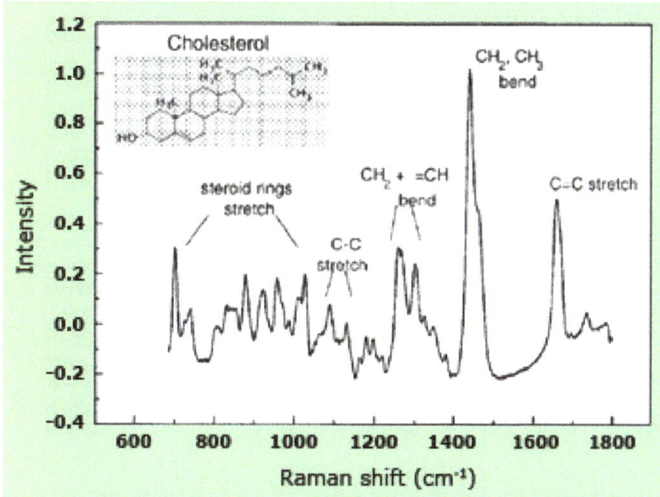

As shown in the figure below, the Raman and Surface Enhanced Raman Spectra can be exploited for biosensors.

1. The Raman spectrum of DTNB (1 mM DTNB in d2H2O);
2. SERS spectrum of BSA-IgG-AuNP;
3. SERS spectrum of TNB-BSAIgG- AuNP. Radiation of 633 nm was used for excitation.

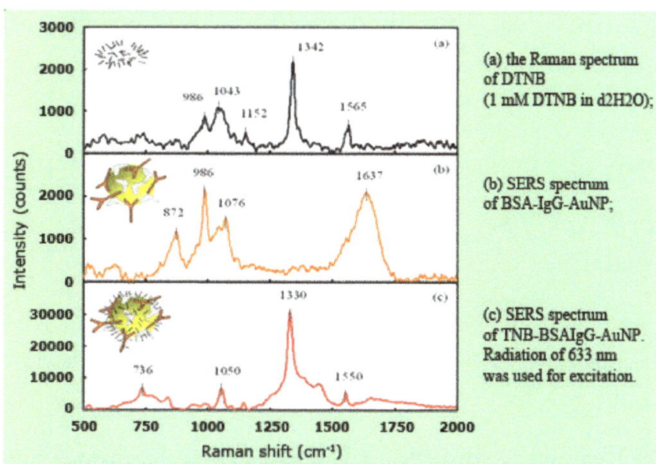

Nuclear Magnetic Resonance (NMR) / Magnetic Resonance Imaging (MRI)

A nucleus with non-zero spin has angular momentum and a magnetic moment such as 1H, 31P, 13C, 19F

This nuclear magnetic moment interacts with the external magnetic field.

In resonance condition, this absorbs the energy.

To obtain the image: vary the external field as a function of position to localize signals.

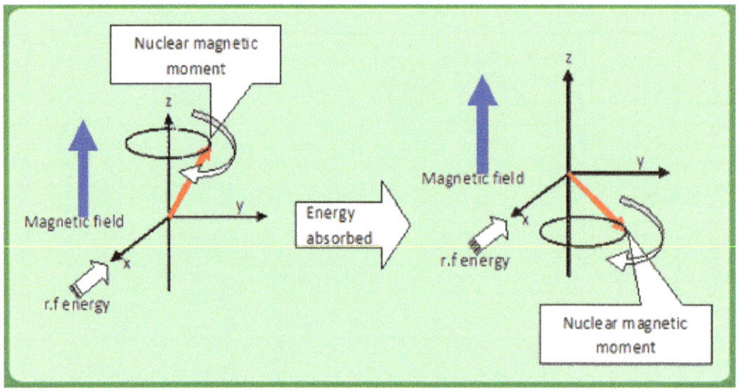

This method is used to acquire the image inside the body especially inside the brain. Having the image from different angles, the medical problems are located.

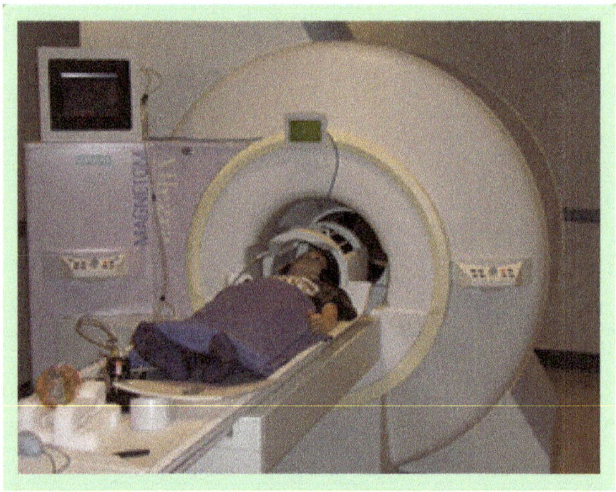

"Matter is made of fundamental building blocks", idea came from ancient Greeks.

They speculated that *earth, air, fire,* and *water* might form the basic elements from which the physical world is constructed.

They also developed various schools of thought about the ultimate nature of matter. Perhaps the most remarkable was the atomist school founded by the ancient Greeks Leucippus of Miletus and Democritus of Thrace about 440 bc.

For purely philosophical reasons, and without benefit of experimental evidence, they developed the notion that matter consists of indivisible and indestructible atoms.

Today, we have come a long way developing many thoughts and ideas, carrying out several experimentations and producing lot of evidences.

An understanding of the structure of atoms and molecules is therefore a prerequisite for entering into several of the areas discussed earlier.

The basic atomic and molecular physics is the scientific study of the structure of the atoms and molecules, its energy states, and its interactions with other particles and with electric and magnetic fields.

References

- Fewell, M. P. (1995). "The atomic nuclide with the highest mean binding energy". American Journal of Physics. 63 (7): 653–658. Bibcode:1995AmJPh..63..653F. doi:10.1119/1.17828
- Knauth, D. C.; Knauth, D. C.; Lambert, David L.; Crane, P. (2000). "Newly synthesized lithium in the interstellar medium". Nature. 405 (6787): 656–58. doi:10.1038/35015028. PMID 10864316
- Liang, Z.-P.; Haacke, E. M. (1999). Webster, J. G., ed. Encyclopedia of Electrical and Electronics Engineering: Magnetic Resonance Imaging. vol. 2. John Wiley & Sons. pp. 412–426. ISBN 0-471-13946-7
- Rutherford, E. (1911). "The Scattering of α and β Particles by Matter and the Structure of the Atom" (PDF). Philosophical Magazine. 21 (125): 669–88. doi:10.1080/14786440508637080
- Campbell, Neil A.; Brad Williamson; Robin J. Heyden (2006). Biology: Exploring Life. Boston, Massachusetts: Pearson Prentice Hall. ISBN 0-13-250882-6. Retrieved 2012-02-05
- Makhijani, Arjun; Saleska, Scott (2 March 2001). "Basics of Nuclear Physics and Fission". Institute for Energy and Environmental Research. Archived from the original on 16 January 2007. Retrieved 3 January 2007
- Roger L. DeKock; Harry B. Gray; Harry B. Gray (1989). Chemical structure and bonding. University Science Books. p. 199. ISBN 0-935702-61-X
- Matis, Howard S. (9 August 2000). "The Isotopes of Hydrogen". Guide to the Nuclear Wall Chart. Lawrence Berkeley National Lab. Archived from the original on 18 December 2007. Retrieved 21 December 2007
- Wink, Donald J.; Fetzer-Gislason, Sharon; McNicholas, Sheila (2003-03-01). The Practice of Chemistry. Macmillan. ISBN 9780716748717
- "Atomic Weights and Isotopic Compositions for All Elements". National Institute of Standards and Technology. Archived from the original on 31 December 2006. Retrieved 4 January 2007
- Meitner, Lise; Frisch, Otto Robert (1939). "Disintegration of uranium by neutrons: a new type of nuclear reaction". Nature. 143 (3615): 239–240. Bibcode:1939Natur.143..239M. doi:10.1038/143239a0
- I. R. Kenyon (2008). The Light Fantastic – Introduction to Classic and Quantum Optics. Oxford University Press. ISBN 978-0-19-856646-5
- R. Eisberg; R. Resnick (1985). Quantum Physics of Atoms, Molecules, Solids, Nuclei, and Particles (2nd ed.). John Wiley & Sons. ISBN 978-0-471-87373-0
- Staff (9 October 2001). "Bose-Einstein Condensate: A New Form of Matter". National Institute of Standards and Technology. Archived from the original on 3 January 2008. Retrieved 16 January 2008
- P.W. Atkins (1977). Molecular Quantum Mechanics Parts I and II: An Introduction to QUANTUM CHEMISTRY (Volume 1). Oxford University Press. ISBN 0-19-855129-0
- Colton, Imogen; Fyffe, Jeanette (3 February 1999). "Super Atoms from Bose-Einstein Condensation". The University of Melbourne. Archived from the original on 29 August 2007. Retrieved 6 February 2008
- J. R. Hook; H. E. Hall (2010). Solid State Physics (2nd ed.). Manchester Physics Series, John Wiley & Sons. ISBN 978-0-471-92804-1
- Jacox, Marilyn; Gadzuk, J. William (November 1997). "Scanning Tunneling Microscope". National Institute of Standards and Technology. Archived from the original on 7 January 2008. Retrieved 11 January 2008
- Jacoby, Mitch (2006). "As-yet-unsynthesized superheavy atom should form a stable diatomic molecule with fluorine". Chemical & Engineering News. 84 (10): 19. doi:10.1021/cen-v084n010.p019a
- Scully, Marlan O.; Lamb, Willis E.; Barut, Asim (1987). "On the theory of the Stern-Gerlach apparatus". Foundations of Physics. 17 (6): 575–583. Bibcode:1987FoPh...17..575S. doi:10.1007/BF01882788

Electromagnetic Radiation: Quantization of Matter and Light

2

Classical mechanics could not explain many experiments such as the photoelectric effect, Compton effect and proposals by Bohr and Rutherford. It was only with the development of quantum mechanics that many of these theories could be satisfactorily understood. This chapter focuses on Maxwell's equations, electromagnetic radiation, wave-particle duality, radiative transition, spectral broadening, etc.

Classical Physics

i. The events around us can be observed and directly measured with instruments.

ii. There is a close link between our perception and the explanation.

iii. For any dynamical physical system its state can be determined with the knowledge of its initial state.

iv. The matter (macroscopic quantity) is treated as localized entity with the state having defined positions and velocities at any instant of time. So, the matter is having "particle nature" and follows the Newtonian Mechanics.

v. Radiation/light is treated as classical electromagnetic wave developed by Maxwell. This is the kind of wave nature of light.

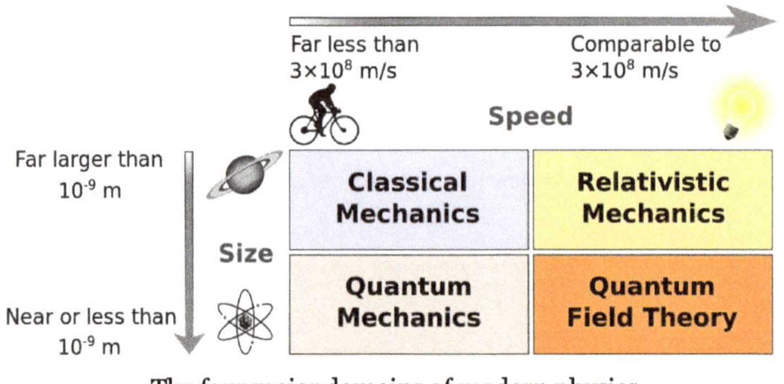

The four major domains of modern physics

Classical physics refers to theories of physics that predate modern, more complete, or more widely applicable theories. If a currently accepted theory is considered to be "modern," and its introduction represented a major paradigm shift, then the previous theories, or new theories based on the older paradigm, will often be referred to as belonging to the realm of "classical" physics.

As such, the definition of a classical theory depends on context. Classical physical concepts are often used when modern theories are unnecessarily complex for a particular situation.

Overview

Classical theory has at least two distinct meanings in physics. In the context of quantum mechanics, classical theory refers to theories of physics that do not use the quantisation paradigm, which includes classical mechanics and relativity. Likewise, classical field theories, such as general relativity and classical electromagnetism, are those that do not use quantum mechanics. In the context of general and special relativity, classical theories are those that obey Galilean relativity.

Among the branches of theory included in classical physics are:

- Classical mechanics
    - Newton's laws of motion
    - Classical Lagrangian and Hamiltonian formalisms
- Classical electrodynamics (Maxwell's Equations)
- Classical thermodynamics
- Special relativity and general relativity
- Classical chaos theory and nonlinear dynamics

Comparison with Modern Physics

In contrast to classical physics, "modern physics" is a slightly looser term which may refer to just quantum physics or to 20th and 21st century physics in general. Modern physics includes quantum theory and relativity, when applicable.

A physical system can be described by classical physics when it satisfies conditions such that the laws of classical physics are approximately valid. In practice, physical objects ranging from those larger than atoms and molecules, to objects in the macroscopic and astronomical realm, can be well-described (understood) with classical mechanics. Beginning at the atomic level and lower, the laws of classical physics break down and generally do not provide a correct description of nature. Electromagnetic fields and forces can be described well by classical electrodynamics at length scales and field strengths large enough that quantum mechanical effects are negligible. Unlike quantum physics, classical physics is generally characterized by the principle of complete determinism, although deterministic interpretations of quantum mechanics do exist.

From the point of view of classical physics as being non-relativistic physics, the predictions of general and special relativity are significantly different than those of classical theories, particularly concerning the passage of time, the geometry of space, the motion of bodies in free fall, and the propagation of light. Traditionally, light was reconciled with classical mechanics by assuming the existence of a stationary medium through which light propagated, the luminiferous aether, which was later shown not to exist.

Mathematically, classical physics equations are those in which Planck's constant does not appear.

According to the correspondence principle and Ehrenfest's theorem, as a system becomes larger or more massive the classical dynamics tends to emerge, with some exceptions, such as superfluidity. This is why we can usually ignore quantum mechanics when dealing with everyday objects and the classical description will suffice. However, one of the most vigorous on-going fields of research in physics is classical-quantum correspondence. This field of research is concerned with the discovery of how the laws of quantum physics give rise to classical physics found at the limit of the large scales of the classical level.

Computer Modeling and Manual Calculation, Modern and Classic Comparison

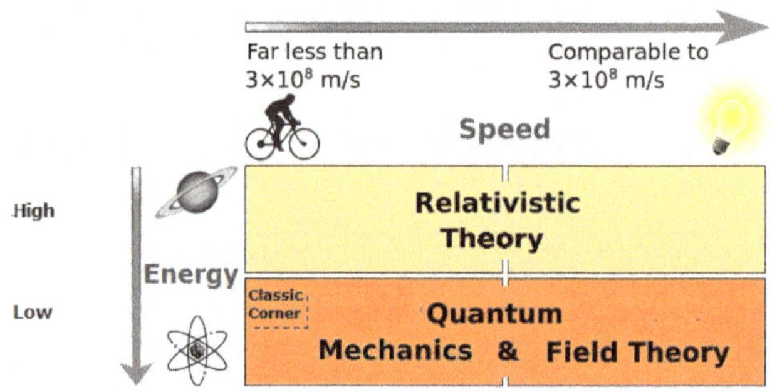

A computer model would use quantum theory and relativistic theory only

Today a computer performs millions of arithmetic operations in seconds to solve a classical differential equation, while Newton (one of the fathers of the differential calculus) would take hours to solve the same equation by manual calculation, even if he were the discoverer of that particular equation.

Computer modeling is essential for quantum and relativistic physics. Classic physics is considered the limit of quantum mechanics for large number of particles. On the other hand, classic mechanics is derived from relativistic mechanics. For example, in many formulations from special relativity, a correction factor $(v/c)^2$ appears, where v is the velocity of the object and c is the speed of light. For velocities much smaller than that of light, one can neglect the terms with c^2 and higher that appear. These formulas then reduce to the standard definitions of Newtonian kinetic energy and momentum. This is as it should be, for special relativity must agree with Newtonian mechanics at low velocities. Computer modeling has to be as real as possible. Classical physics would introduce an error as in the superfluidity case. In order to produce reliable models of the world, we can not use classic physics. It is true that quantum theories consume time and computer resources, and the equations of classical physics could be resorted to provide a quick solution, but such a solution would lack reliability.

Computer modeling would use only the energy criteria to determine which theory to use: relativity or quantum theory, when attempting to describe the behavior of an object. A physicist would use a classical model to provide an approximation before more exacting models are applied and those calculations proceed.

In a computer model, there is no need to use the speed of the object if classical physics is excluded. Low energy objects would be handled by quantum theory and high energy objects by relativity theory.

Maxwell's Equations

Maxwell's equations (mid-left) as featured on a monument in front of Warsaw University's Centre of New Technologies

Maxwell's equations are a set of partial differential equations that, together with the Lorentz force law, form the foundation of classical electromagnetism, classical optics, and electric circuits. They underpin all electric, optical and radio technologies such as power generation, electric motors, wireless communication, cameras, televisions, computers etc. Maxwell's equations describe how electric and magnetic fields are generated by charges, currents and changes of each other. One important consequence of the equations is the demonstration of how fluctuating electric and magnetic fields can propagate at the speed of light. This electromagnetic radiation manifests itself in manifold ways from radio waves to light and X- or γ-rays. The equations are named after the physicist and mathematician James Clerk Maxwell, who between 1861 and 1862 published an early form of the equations, and first proposed that light is an electromagnetic phenomenon.

The equations have two major variants. The microscopic Maxwell equations have universal applicability but are unwieldy for common calculations. They relate the electric and magnetic fields to total charge and total current, including the complicated charges and currents in materials at the atomic scale. The "macroscopic" Maxwell equations define two new auxiliary fields that describe the large-scale behaviour of matter without having to consider atomic scale details. However, their use requires experimentally determining parameters for a phenomenological description of the electromagnetic response of materials.

The term "Maxwell's equations" is often used for equivalent alternative formulations. Versions of Maxwell's equations based on the electric and magnetic potentials are preferred for explicitly solving the equations as a boundary value problem, analytical mechanics, or for use in quantum mechanics. The space-time formulations (i.e., on space-time rather than space and time separately), are commonly used in high energy and gravitational physics because they make the compatibility of the equations with special and general relativity manifest. In fact, historically, Einstein developed special and general relativity to accommodate the absolute speed of light that drops out of the Maxwell equations with the principle that only relative movement has physical consequences.

Since the mid-20th century, it has been understood that Maxwell's equations are not exact but are a classical field theory approximation of some aspects of the fundamental theory of quantum electrodynamics, although some quantum features, such as quantum entanglement, are completely absent and in no way approximated. (For example, quantum cryptography has no approximate version in Maxwell theory.) In many situations, though, deviations from Maxwell's equations are immeasurably small. Exceptions include nonclassical light, photon-photon scattering, quantum optics, and many other phenomena related to photons or virtual photons.

Formulation in Terms of Electric and Magnetic Fields (Microscopic or in Vacuum Version)

In the electric and magnetic field formulation there are four equations. The two inhomogeneous equations describe how the fields vary in space due to sources. Gauss's law describes how electric fields emanate from electric charges. Gauss's law for magnetism describes magnetic fields as closed field lines *not due to magnetic monopoles*. The two homogeneous equations describe how the fields "circulate" around their respective sources. Ampère's law with Maxwell's addition describes how the magnetic field "circulates" around electric currents and time varying electric fields, while Faraday's law describes how the electric field "circulates" around time varying magnetic fields.

A separate law of nature, the Lorentz force law, describes how the electric and magnetic field act on charged particles and currents. A version of this law was included in the original equations by Maxwell but, by convention, is no longer.

The precise formulation of Maxwell's equations depends on the precise definition of the quantities involved. Conventions differ with the unit systems, because various definitions and dimensions are changed by absorbing dimensionful factors like the speed of light *c*. This makes constants come out differently. The most common form is based on conventions used when quantities measured using SI units, but other commonly used conventions are used with other units including Gaussian units based on the cgs system, Lorentz–Heaviside units (used mainly in particle physics), and Planck units (used in theoretical physics).

The vector calculus formulation below has become standard. It is mathematically much more convenient than Maxwell's original 20 equations and is due to Oliver Heaviside The differential and integral equations formulations are mathematically equivalent and are both useful. The integral formulation relates fields within a region of space to fields on the boundary and can often be used to simplify and directly calculate fields from symmetric distributions of charges and currents. On the other hand, the differential equations are purely *local* and are a more natural starting point for calculating the fields in more complicated (less symmetric) situations, for example using finite element analysis.

Formulation in SI Units Convention

Name	Integral equations	Differential equations	Meaning
Gauss's law	$\oiint_{\partial\Omega} \mathbf{E}\cdot d\mathbf{S} = \frac{1}{\varepsilon_0}\iiint_{\Omega} \rho \, dV$	$\nabla\cdot\mathbf{E} = \frac{\rho}{\varepsilon_0}$	The electric flux leaving a volume is proportional to the charge inside.
Gauss's law for magne-tism	$\oiint_{\partial\Omega} \mathbf{B}\cdot d\mathbf{S} = 0$	$\nabla\cdot\mathbf{B} = 0$	There are no magnetic monopoles; the total magnetic flux through a closed sur-face is zero.
Max-well–Faraday equation (Fara-day's law of induc-tion)	$\oint_{\partial\Sigma} \mathbf{E}\cdot d\ell = -\frac{d}{dt}\iint_{\Sigma} \mathbf{B}\cdot d\mathbf{S}$	$\nabla\times\mathbf{E} = -\frac{\partial\mathbf{B}}{\partial t}$	The voltage induced in a closed circuit is proportional to the rate of change of the magnetic flux it enclos-es.
Ampère's circuital law (with Max-well's addition)	$\oint_{\partial\Sigma} \mathbf{B}\cdot d\ell = \mu_0\iint_{\Sigma} \mathbf{J}\cdot d\mathbf{S} + \mu_0\varepsilon_0\frac{d}{dt}\iint_{\Sigma} \mathbf{E}\cdot d\mathbf{S}$	$\nabla\times\mathbf{B} = \mu_0\left(\mathbf{J} + \varepsilon_0\frac{\partial\mathbf{E}}{\partial t}\right)$	The mag-netic field induced around a closed loop is propor-tional to the electric current plus displacement current (rate of change of electric field) it encloses.

Formulation in Gaussian Units Convention

Gaussian units are a popular system of units, that are part of the centimetre–gram–second system of units (cgs). When using cgs units it is conventional to use a slightly different definition of electric field $\mathbf{E}_{cgs} = c^{-1}\,\mathbf{E}_{SI}$. This implies that the modified electric and magnetic field have the same units (in the SI convention this is not the case making dimensional analysis of the equations different: e.g. for an electromagnetic wave in vacuum $|\mathbf{E}|_{SI} = |c|_{SI} |\mathbf{B}|_{SI}$,).

The CGS system uses a unit of charge defined in such a way that the permittivity of the vacuum $\varepsilon_0 = \frac{1}{4\pi c}$,hence $\mu_0 = \frac{4\pi}{c}$. These units are sometimes preferred over SI units in the context of special relativity, since when using them, the components of the electromagnetic tensor, the Lorentz covariant object describing the electromagnetic field, have the same unit without constant factors. Using these different conventions, the Maxwell equations become:

Equations in Gaussian units convention			
Name	**Integral equations**	**Differential equations**	**Meaning**
Gauss's law	$\oiint_{\partial\Omega} \mathbf{E}\cdot d\mathbf{S} = 4\pi \iiint_{\Omega} \rho dV$	$\nabla\cdot\mathbf{E} = 4\pi\rho$	The electric flux leaving a volume is proportional to the charge inside.
Gauss's law for magnetism	$\oiint_{\partial\Omega} \mathbf{B}\cdot d\mathbf{S} = 0$	$\nabla\cdot\mathbf{B} = 0$	There are no magnetic monopoles; the total magnetic flux through a closed surface is zero.
Maxwell–Faraday equation (Faraday's law of induction)	$\oint_{\partial\Sigma} \mathbf{E}\cdot d\ell = -\frac{1}{c}\frac{d}{dt}\iint_{\Sigma} \mathbf{B}\cdot d\mathbf{S}$	$\nabla\times\mathbf{E} = -\frac{1}{c}\frac{\partial\mathbf{B}}{\partial t}$	The voltage induced in a closed circuit is proportional to the rate of change of the magnetic flux it encloses.
Ampère's law (with Maxwell's extension)	$\oint_{\partial\Sigma} \mathbf{B}\cdot d\ell = \frac{1}{c}\left(4\pi\iint_{\Sigma} \mathbf{J}\cdot d\mathbf{S} + \frac{d}{dt}\iint_{\Sigma} \mathbf{E}\cdot d\mathbf{S}\right)$	$\nabla\times\mathbf{B} = \frac{1}{c}\left(4\pi\mathbf{J} + \frac{\partial\mathbf{E}}{\partial t}\right)$	The magnetic field integrated around a closed loop is proportional to the electric current plus displacement current (rate of change of electric field) it encloses.

Key to the Notation

Symbols in bold represent vector quantities, and symbols in *italics* represent scalar quantities, unless otherwise indicated.

The equations introduce the electric field, E, a vector field, and the magnetic field, B, a pseudovector field, each generally having a time and location dependence. The sources are

- the electric charge density (charge per unit volume), ρ, and
- the electric current density (current per unit area), **J**.

The universal constants appearing in the equations are

- the permittivity of free space, ε_0, and
- the permeability of free space, μ_0.

Differential Equations

In the differential equations,

- the nabla symbol, ∇, denotes the three-dimensional gradient operator,

- the $\nabla\cdot$ symbol denotes the divergence operator,
- the $\nabla\times$ symbol denotes the curl operator.

Integral Equations

In the integral equations,

- Ω is any fixed volume with closed boundary surface $\partial\Omega$, and
- Σ is any fixed surface with closed boundary curve $\partial\Sigma$,

Here a *fixed* volume or surface means that it does not change over time. The equations are correct, complete and a little easier to interpret with time-independent surfaces. However, since the surface is time-independent, we can bring the differentiation under the integral sign in Faraday's law:

$$\frac{d}{dt}\iint_{\Sigma}\mathbf{B}\cdot d\mathbf{S}=\iint_{\Sigma}\frac{\partial\mathbf{B}}{\partial t}\cdot d\mathbf{S},$$

The Maxwell's equations can be formulated with possibly time dependent surfaces and volumes by substituting the lefthand side with the righthand side in the integral equation version of the Maxwell equations.

- $\oiint_{\partial\Omega}$ is a surface integral over the surface $\partial\Omega$, (the loop indicates that the boundary surface is closed)
- $\iiint_{\Omega}$ is a volume integral over the volume Ω,
- $\oint_{\partial\Sigma}$ is a surface integral over the surface Σ,
- $\iint_{\Sigma}$ is a line integral around the curve $\partial\Sigma$ (the loop indicates that the boundary curve is closed).
- The volume integral over Ω of the total charge density ρ, is the *total* electric charge Q contained in Ω:

  $$Q=\iiint_{\Omega}\rho dV,$$

  where dV is the volume element.
- The *net* electric current I is the surface integral of the electric current density **J** passing through a fixed surface, Σ:

  $$I=\iint_{\Sigma}\mathbf{J}\cdot d\mathbf{S},$$

  where d**S** denotes the vector element of surface area, S, normal to surface, Σ. (Vector area is also denoted by A rather than S, but this conflicts with the magnetic potential, a separate vector field).

Relationship Between Differential and Integral Formulations

The equivalence of the differential and integral formulations are a consequence of the Gauss divergence theorem and the Kelvin–Stokes theorem.

Flux and Divergence

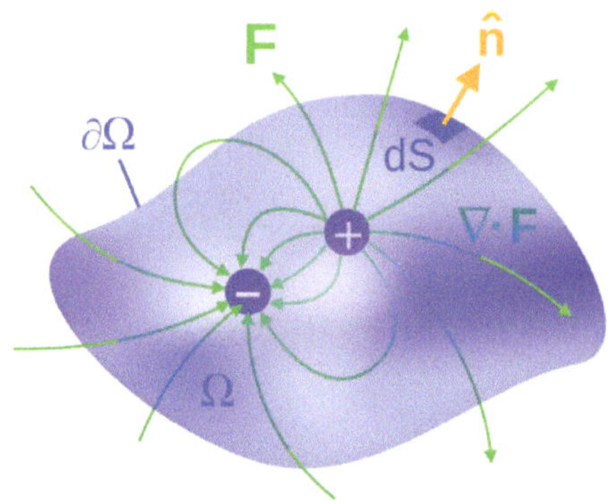

Volume Ω and its closed boundary $\partial\Omega$, containing (respectively enclosing) a source (+) and sink (−) of a vector field F. Here, F could be the E field with source electric charges, but not the B field which has no magnetic charges as shown. The outward unit normal is n

The "sources of the fields" (i.e. their divergence) can be determined from the surface integrals of the fields through the closed surface $\partial\Omega$. E.g. the electric flux is

$$\oiint_{\partial\Omega} \mathbf{E}\cdot \mathrm{d}\mathbf{S} = \iiint_{\Omega} \nabla\cdot\mathbf{E}\,\mathrm{d}V$$

where the last equality uses the Gauss divergence theorem. Using the integral version of Gauss's equation we can rewrite this to

$$\iiint_{\Omega} (\nabla\cdot\mathbf{E} - \frac{\rho}{\epsilon_0})\mathrm{d}V = 0$$

Since Ω can be chosen arbitrarily, e.g. as an arbitrary small ball with arbitrary center, this implies that the integrand must be zero, which is the differential equations formulation of Gauss equation up to a trivial rearrangement. Gauss's law for magnetism in differential equations form follows likewise from the integral form by rewriting the magnetic flux

$$\oiint_{\partial\Omega} \mathbf{B}\cdot \mathrm{d}\mathbf{S} = \iiint_{\Omega} \nabla\cdot\mathbf{B}\,\mathrm{d}V = 0.$$

Circulation and Curl

The "circulation of the fields" (i.e. their curls) can be determined from the line integrals of the fields around the closed curve $\partial\Sigma$. E.g. for the magnetic field

$$\oint_{\partial\Sigma} \mathbf{B}\cdot \mathrm{d}\ell = \iint_{\Sigma} (\nabla\times\mathbf{B})\cdot \mathrm{d}\mathbf{S},$$

where we used the Kelvin-Stokes theorem. Using the modified Ampere law in integral form and the writing the time derivative of the flux as the surface integral of the partial time derivative of **E** we conclude that

$$\iint_{\Sigma}\left(\nabla\times\mathbf{B}-\mu_0(\mathbf{J}+\epsilon_0\frac{\partial\mathbf{E}}{\partial t})\right)\cdot d\mathbf{S}=0$$

Since Σ can be chosen arbitrarily, e.g. as an arbitrary small, arbitrary oriented, and arbitrary centered disk, we conclude that the integrand must be zero. This is Ampere's modified law in differential equations form up to a trivial rearrangement. Likewise, the Faraday law in differential equations form follows from rewriting the integral form using the Kelvin-Stokes theorem.

The line integrals and curls are analogous to quantities in classical fluid dynamics: the circulation of a fluid is the line integral of the fluid's flow velocity field around a closed loop, and the vorticity of the fluid is the curl of the velocity field.

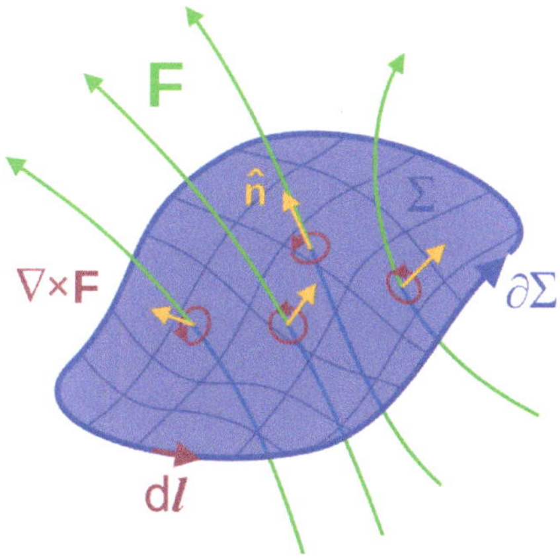

Surface Σ with closed boundary ∂Σ. F could be the E or B fields. Again, n is the unit normal.
(The curl of a vector field doesn't literally look like the "circulations", this is a heuristic depiction)

Conceptual Descriptions

Gauss's Law

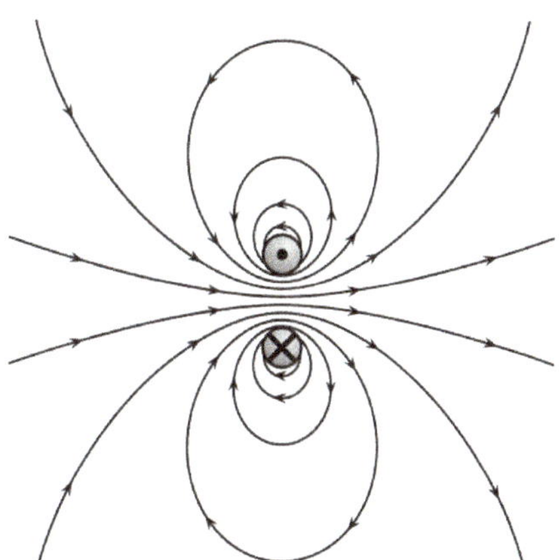

Gauss's law for magnetism: magnetic field lines never begin nor end but form loops or extend to infinity as shown here with the magnetic field due to a ring of current

Gauss's law describes the relationship between a static electric field and the electric charges that cause it: The static electric field points away from positive charges and towards negative charges. In the field line description, electric field lines begin only at positive electric charges and end only at negative electric charges. 'Counting' the number of field lines passing through a closed surface, therefore, yields the total charge (including bound charge due to polarization of material) enclosed by that surface divided by dielectricity of free space (the vacuum permittivity). More technically, it relates the electric flux through any hypothetical closed "Gaussian surface" to the enclosed electric charge.

Gauss's Law for Magnetism

Gauss's law for magnetism states that there are no "magnetic charges" (also called magnetic monopoles), analogous to electric charges. Instead, the magnetic field due to materials is generated by a configuration called a dipole. Magnetic dipoles are best represented as loops of current but resemble positive and negative 'magnetic charges', inseparably bound together, having no net 'magnetic charge'. In terms of field lines, this equation states that magnetic field lines neither begin nor end but make loops or extend to infinity and back. In other words, any magnetic field line that enters a given volume must somewhere exit that volume. Equivalent technical statements are that the sum total magnetic flux through any Gaussian surface is zero, or that the magnetic field is a solenoidal vector field.

Faraday's Law

In a geomagnetic storm, a surge in the flux of charged particles temporarily alters Earth's magnetic field, which induces electric fields in Earth's atmosphere, thus causing surges in electrical power grids. Artist's rendition; sizes are not to scale

The Maxwell-Faraday's equation version of Faraday's law describes how a time varying magnetic field creates ("induces") an electric field. This dynamically induced electric field has closed field lines just as the magnetic field, if not superposed by a static (charge induced) electric field. This aspect of electromagnetic induction is the operating principle behind many electric generators: for example, a rotating bar magnet creates a changing magnetic field, which in turn generates an electric field in a nearby wire.

Ampère's Law with Maxwell's Addition

Ampère's law with Maxwell's addition states that magnetic fields can be generated in two ways: by electric current (this was the original "Ampère's law") and by changing electric fields (this was "Maxwell's addition").

Maxwell's addition to Ampère's law is particularly important: it makes the set of equations mathematically consistent for non static fields, without changing the laws of Ampere and Gauss for static fields. However, as a consequence, it predicts that a changing magnetic field induces an electric field and vice versa. Therefore, these equations allow self-sustaining "electromagnetic waves" to travel through empty space.

Magnetic core memory (1954) is an application of Ampère's law. Each core stores one bit of data

The speed calculated for electromagnetic waves, which could be predicted from experiments on charges and currents, exactly matches the speed of light; indeed, light *is* one form of electromagnetic radiation (as are X-rays, radio waves, and others). Maxwell understood the connection between electromagnetic waves and light in 1861, thereby unifying the theories of electromagnetism and optics.

Charge Conservation

The invariance of charge can be derived as a corollary of Maxwell's equations. The left hand side of the modified Ampere's Law has zero divergence by the div-curl-identity. Combining the right hand side, Gauss's law, and interchange of derivatives gives:

$$0 = \nabla \cdot \nabla \times \mathbf{B} = \mu_0 \left(\nabla \cdot \mathbf{J} + \varepsilon_0 \frac{\partial}{\partial t} \nabla \cdot \mathbf{E} \right) = \mu_0 \left(\nabla \cdot \mathbf{J} + \frac{\partial \rho}{\partial t} \right)$$

i.e.

$$\frac{\partial \rho}{\partial t} + \nabla \cdot \mathbf{J} = 0$$

By the Gauss divergence theorem that means that the rate of change of the charge in a fixed volume equals the current flowing in or out of the boundary

$$\frac{d}{dt} Q_\Omega = \frac{d}{dt} \iiint_\Omega \rho \mathrm{d}V = -\oint\!\!\!\oint_{\partial\Omega} \mathbf{J} \cdot \mathrm{d}\mathbf{S} = -I_{\partial\Omega}.$$

In particular, in an isolated system the total charge is conserved.

Vacuum Equations, Electromagnetic Waves and Speed of Light

In a region with no charges ($\rho = 0$) and no currents ($\mathbf{J} = \mathbf{0}$), such as in a vacuum, Maxwell's equations reduce to:

$$\nabla \cdot \mathbf{E} = 0 \quad \nabla \times \mathbf{E} = -\frac{\partial \mathbf{B}}{\partial t},$$
$$\nabla \cdot \mathbf{B} = 0 \quad \nabla \times \mathbf{B} = \frac{1}{c^2}\frac{\partial \mathbf{E}}{\partial t}.$$

Taking the curl ($\nabla\times$) of the curl equations, and using the curl of the curl identity $\nabla \times (\nabla \times \mathbf{X}) = \nabla(\nabla\cdot\mathbf{X}) - \nabla^2\mathbf{X}$ we obtain the wave equations

$$\frac{1}{c^2}\frac{\partial^2 \mathbf{E}}{\partial t^2} - \nabla^2 \mathbf{E} = 0$$
$$\frac{1}{c^2}\frac{\partial^2 \mathbf{B}}{\partial t^2} - \nabla^2 \mathbf{B} = 0$$

which identify

$$c = \frac{1}{\sqrt{\mu_0 \varepsilon_0}} = 2.99792458 \times 10^8 \,\mathrm{ms}^{-1}$$

with the speed of light in free space. In materials with relative permittivity, ε_r, and relative permeability, μ_r, the phase velocity of light becomes

$$v_p = \frac{1}{\sqrt{\mu_0 \mu_r \varepsilon_0 \varepsilon_r}}$$

which is usually less than c.

In addition, E and B are mutually perpendicular to each other and the direction of wave propagation, and are in phase with each other. A sinusoidal plane wave is one special solution of these equations. Maxwell's equations explain how these waves can physically propagate through space. The changing magnetic field creates a changing electric field through Faraday's law. In turn, that electric field creates a changing magnetic field through Maxwell's addition to Ampère's law. This perpetual cycle allows these waves, now known as electromagnetic radiation, to move through space at velocity c.

Macroscopic Formulation

The *microscopic* variant of Maxwell's equation is the version given above. It expresses the electric **E** field and the magnetic B field in terms of the *total charge* and *total current* present, including the charges and currents at the atomic level. The "microscopic" form is sometimes called the "general" form of Maxwell's equations. The macroscopic variant of Maxwell's equation is equally general, however, with the difference being one of bookkeeping.

The "microscopic" variant is sometimes called "Maxwell's equations in a vacuum". This refers to the fact that the material medium is not built into the structure of the equation; it does not mean that space is empty of charge or current. They are also known as the "Maxwell-Lorentz equations". Lorentz tried to use these equations to predict the macroscopic properties of bulk matter from the physical behavior of its microscopic constituents.

"Maxwell's macroscopic equations", also known as Maxwell's equations in matter, are more similar to those that Maxwell introduced himself.

Name	Integral equations (SI convention)	Differential equations (SI convention)	Differential equations (Gaussian convention)
Gauss's law	$\oiint_{\partial\Omega} \mathbf{D}\cdot d\mathbf{S} = \iiint_{\Omega} \rho_f \, dV$	$\nabla\cdot\mathbf{D} = \rho_f$	$\nabla\cdot\mathbf{D} = 4\pi\rho_f$
Gauss's law for magnetism	$\oiint_{\partial\Omega} \mathbf{B}\cdot d\mathbf{S} = 0$	$\nabla\cdot\mathbf{B} = 0$	$\nabla\cdot\mathbf{B} = 0$
Maxwell–Faraday equation (Faraday's law of induction)	$\oint_{\partial\Sigma} \mathbf{E}\cdot d\ell = -\frac{d}{dt}\iint_{\Sigma} \mathbf{B}\cdot d\mathbf{S}$	$\nabla\times\mathbf{E} = -\frac{\partial\mathbf{B}}{\partial t}$	$\nabla\times\mathbf{E} = -\frac{1}{c}\frac{\partial\mathbf{B}}{\partial t}$
Ampère's circuital law (with Maxwell's addition)	$\oint_{\partial\Sigma} \mathbf{H}\cdot d\ell = \iint_{\Sigma} \mathbf{J}_f\cdot d\mathbf{S} + \frac{d}{dt}\iint_{\Sigma} \mathbf{D}\cdot d\mathbf{S}$	$\nabla\times\mathbf{H} = \mathbf{J}_f + \frac{\partial\mathbf{D}}{\partial t}$	$\nabla\times\mathbf{H} = \frac{1}{c}\left(4\pi\mathbf{J}_f + \frac{\partial\mathbf{D}}{\partial t}\right)$

Unlike the "microscopic" equations, the "macroscopic" equations separate out the bound charge Q_b and bound current I_b to obtain equations that depend only on the free charges Q_f and currents I_f. This factorization can be made by splitting the total electric charge and current as follows:

$$Q = Q_f + Q_b = \iiint_{\Omega} (\rho_f + \rho_b)dV = \iiint_{\Omega} \rho dV$$

$$I = I_f + I_b = \iint_{\Sigma} (\mathbf{J}_f + \mathbf{J}_b)\cdot d\mathbf{S} = \iint_{\Sigma} \mathbf{J}\cdot d\mathbf{S}$$

Correspondingly, the total current density J splits into free J_f and bound J_b components, and similarly the total charge density ρ splits into free ρ_f and bound ρ_b parts.

The cost of this factorization is that additional fields, the displacement field D and the magnetizing field H, are defined and need to be determined. Phenomenological constituent equations relate the additional fields to the electric field E and the magnetic B-field, often through a simple linear relation.

For a detailed description of the differences between the microscopic (*total* charge and current including material contributes or in air/vacuum) and macroscopic (*free* charge and current; practical to use on materials) variants of Maxwell's equations.

Bound Charge and Current

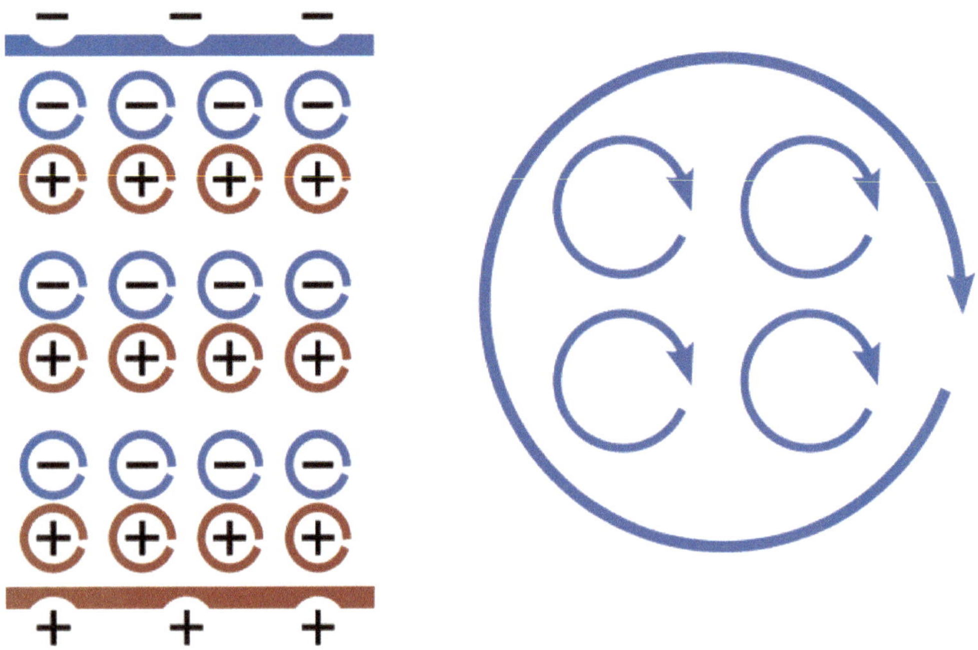

Left: A schematic view of how an assembly of microscopic dipoles produces opposite surface charges as shown at top and bottom. *Right:* How an assembly of microscopic current loops add together to produce a macroscopically circulating current loop. Inside the boundaries, the individual contributions tend to cancel, but at the boundaries no cancelation occurs

When an electric field is applied to a dielectric material its molecules respond by forming microscopic electric dipoles – their atomic nuclei move a tiny distance in the direction of the field, while their electrons move a tiny distance in the opposite direction. This produces a macroscopic bound charge in the material even though all of the charges involved are bound to individual molecules. For example, if every molecule responds the same, similar to that shown in the figure, these tiny movements of charge combine to produce a layer of positive bound charge on one side of the material and a layer of negative charge on the other side. The bound charge is most conveniently described in terms of the polarization P of the material, its dipole moment per unit volume. If P is uniform, a macroscopic separation of charge is produced only at the surfaces where P enters and leaves the material. For non-uniform P, a charge is also produced in the bulk.

Somewhat similarly, in all materials the constituent atoms exhibit magnetic moments that are intrinsically linked to the angular momentum of the components of the atoms, most notably their electrons. The connection to angular momentum suggests the picture of an assembly of microscopic current loops. Outside the material, an assembly of such microscopic current loops is not different from a macroscopic current circulating around the material's surface, despite the fact that no individual charge is traveling a large distance. These bound currents can be described using the magnetization M.

The very complicated and granular bound charges and bound currents, therefore, can be represented on the macroscopic scale in terms of P and M which average these charges and currents on a sufficiently large scale so as not to see the granularity of individual atoms, but also sufficiently small that they vary with location in the material. As such, *Maxwell's macroscopic equations* ignore many details on a fine scale that can be unimportant to understanding matters on a gross scale by calculating fields that are averaged over some suitable volume.

Auxiliary Fields, Polarization and Magnetization

The *definitions* (not constitutive relations) of the auxiliary fields are:

$$\mathbf{D}(\mathbf{r},t) = \varepsilon_0 \mathbf{E}(\mathbf{r},t) + \mathbf{P}(\mathbf{r},t)$$

$$\mathbf{H}(\mathbf{r},t) = \frac{1}{\mu_0}\mathbf{B}(\mathbf{r},t) - \mathbf{M}(\mathbf{r},t)$$

where P is the polarization field and M is the magnetization field which are defined in terms of microscopic bound charges and bound currents respectively. The macroscopic bound charge density ρ_b and bound current density J_b in terms of polarization P and magnetization M are then defined as

$$\rho_b = -\nabla \cdot \mathbf{P}$$

$$\mathbf{J}_b = \nabla \times \mathbf{M} + \frac{\partial \mathbf{P}}{\partial t}$$

If we define the total, bound, and free charge and current density by

$$\rho = \rho_b + \rho_f,$$
$$\mathbf{J} = \mathbf{J}_b + \mathbf{J}_f,$$

and use the defining relations above to eliminate D, and H, the "macroscopic" Maxwell's equations reproduce the "microscopic" equations.

Constitutive Relations

In order to apply 'Maxwell's macroscopic equations', it is necessary to specify the relations between displacement field D and the electric field E, as well as the magnetizing field H and the magnetic field B. Equivalently, we have to specify the dependence of the polarisation P (hence the bound charge) and the magnetisation M (hence the bound current) on the applied electric and magnetic field. The equations specifying this response are called constitutive relations. For real-world materials, the constitutive relations are rarely simple, except approximately, and usually determined by experiment.

For materials without polarisation and magnetisation, the constitutive relations are (by definition)

$$\mathbf{D} = \varepsilon_0 \mathbf{E}, \quad \mathbf{H} = \frac{1}{\mu_0}\mathbf{B}$$

where ε_0 is the permittivity of free space and μ_0 the permeability of free space. Since there is no bound charge, the total and the free charge and current are equal.

An alternative viewpoint on the microscopic equations is that they are the macroscopic equations *together* with the statement that vacuum behaves like a perfect linear "material" without additional polarisation and magnetisation. More generally, for linear materials the constitutive relations are

$$\mathbf{D} = \varepsilon \mathbf{E}, \quad \mathbf{H} = \frac{1}{\mu}\mathbf{B}$$

where ε is the permittivity and μ the permeability of the material. For the displacement field D the linear approximation is usually excellent because for all but the most extreme electric fields or temperatures obtainable in the laboratory (high power pulsed lasers) the interatomic electric fields of materials of the order of 10^{11} V/m are much higher than the external field. For the magnetizing field $\mathbf{H}$, however, the linear approximation can break down in common materials like iron leading to phenomena like hysteresis. Even the linear case can have various complications, however.

- For homogeneous materials, ε and μ are constant throughout the material, while for inhomogeneous materials they depend on location within the material (and perhaps time).
- For isotropic materials, ε and μ are scalars, while for anisotropic materials (e.g. due to crystal structure) they are tensors.
- Materials are generally dispersive, so ε and μ depend on the frequency of any incident EM waves.

Even more generally, in the case of non-linear materials (for example nonlinear optics), D and P are not necessarily proportional to E, similarly H or M is not necessarily proportional to B. In general D and H depend on both E and B, on location and time, and possibly other physical quantities.

In applications one also has to describe how the free currents and charge density behave in terms of E and B possibly coupled to other physical quantities like pressure, and the mass, number density, and velocity of charge-carrying particles. E.g., the original equations given by Maxwell (History of Maxwell's equations) included Ohms law in the form

$$\mathbf{J}_f = \sigma \mathbf{E}.$$

Alternative Formulations

Following is a summary of some of the numerous other ways to write the microscopic Maxwell's equations, showing they can be formulated using different mathematical formalisms. In addition, we formulate the equations using "potentials". Originally they were introduced as a convenient way to solve the homogeneous equations, but it was originally thought that all the observable physics was contained in the electric and magnetic fields (or relativistically, the Faraday tensor). The potentials play a central role in quantum mechanics, however, and act quantum mechanically with observable consequences even when the electric and magnetic fields vanish (Aharonov–Bohm effect). SI units are used throughout.

Formalism	Formulation	Homogeneous equations	Inhomogeneous equations
Vector calculus	**Fields 3D Euclidean space + time**	$\nabla \cdot \mathbf{B} = 0$ $\nabla \times \mathbf{E} + \frac{\partial \mathbf{B}}{\partial t} = 0$	$\nabla \cdot \mathbf{E} = \frac{\rho}{\varepsilon_0}$ $\nabla \times \mathbf{B} - \frac{1}{c^2}\frac{\partial \mathbf{E}}{\partial t} = \mu_0 \mathbf{J}$
	Potentials (any gauge) 3D Euclidean space + time	$\mathbf{B} = \nabla \times \mathbf{A}$ $\mathbf{E} = -\nabla\varphi - \frac{\partial \mathbf{A}}{\partial t}$	$-\nabla^2\varphi - \frac{\partial}{\partial t}(\nabla \cdot \mathbf{A}) = \frac{\rho}{\varepsilon_0}$ $(-\nabla^2 + \frac{1}{c^2}\frac{\partial^2}{\partial t^2})\mathbf{A} + \nabla\left(\nabla \cdot \mathbf{A} + \frac{1}{c^2}\frac{\partial\varphi}{\partial t}\right) = \mu_0 \mathbf{J}$
	Potentials (Lorenz gauge) 3D Euclidean space + time	$\mathbf{B} = \nabla \times \mathbf{A}$ $\mathbf{E} = -\nabla\varphi - \frac{\partial \mathbf{A}}{\partial t}$ $\nabla \cdot \mathbf{A} = -\frac{1}{c^2}\frac{\partial\varphi}{\partial t}$	$(-\nabla^2 + \frac{1}{c^2}\frac{\partial^2}{\partial t^2})\varphi = \frac{\rho}{\varepsilon_0}$ $(-\nabla^2 + \frac{1}{c^2}\frac{\partial^2}{\partial t^2})\mathbf{A} = \mu_0 \mathbf{J}$
Tensor calculus	**Fields space + time spatial metric independent of time**	$\partial_{[i}B_{jk]} =$ $\nabla_{[i}B_{jk]} = 0$ $\partial_{[i}E_{j]} + \frac{\partial B_{ij}}{\partial t} =$ $\nabla_{[i}E_{j]} + \frac{\partial B_{ij}}{\partial t} = 0$	$\frac{1}{\sqrt{h}}\partial_i\sqrt{h}E^i =$ $\nabla_i E^i = \frac{\rho}{\epsilon_0}$ $-\frac{1}{\sqrt{h}}\partial_i\sqrt{h}B^{ij} - \frac{1}{c^2}\frac{\partial}{\partial t}E^j =$ $-\nabla_i B^{ij} - \frac{1}{c^2}\frac{\partial E^j}{\partial t} = \mu_0 J^j$
	Potentials space (with topological restrictions) + time spatial metric independent of time	$B_{ij} = \partial_{[i}A_{j]}$ $= \nabla_{[i}A_{j]}$ $E_i = -\frac{\partial A_i}{\partial t} - \partial_i\phi$ $= -\frac{\partial A_i}{\partial t} - \nabla_i\phi$	$-\frac{1}{\sqrt{h}}\partial_i\sqrt{h}(\partial^i\phi + \frac{\partial A^i}{\partial t}) =$ $-\nabla_i\nabla^i\phi - \frac{\partial}{\partial t}\nabla_i A^i = \frac{\rho}{\epsilon_0}$ $-\frac{1}{\sqrt{h}}\partial_i(\sqrt{h}h^{im}h^{jn}\partial_{[m}A_{n]}) + \frac{1}{c^2}\frac{\partial}{\partial t}(\frac{\partial A^j}{\partial t} + \partial^j\phi) =$ $-\nabla_i\nabla^i A^j + \frac{1}{c^2}\frac{\partial^2 A^j}{\partial t^2} + R^j_i A^i + \nabla^j(\nabla_i A^i + \frac{1}{c^2}\frac{\partial\phi}{\partial t}) = \mu_0 J^j$
	Potentials (Lorenz gauge) space (with topological restrictions) + time spatial metric independent of time	$B_{ij} = \partial_{[i}A_{j]}$ $= \nabla_{[i}A_{j]}$ $E_i = -\frac{\partial A_i}{\partial t} - \partial_i\phi$ $= -\frac{\partial A_i}{\partial t} - \nabla_i\phi$ $\nabla_i A^i = -\frac{1}{c^2}\frac{\partial\phi}{\partial t}$	$-\nabla_i\nabla^i\phi + \frac{1}{c^2}\frac{\partial^2\phi}{\partial t^2} = \frac{\rho}{\epsilon_0}$ $-\nabla_i\nabla^i A^j + \frac{1}{c^2}\frac{\partial^2 A^j}{\partial t^2} + R^j_i A^i = \mu_0 J^j$

	Fields **Any space + time**	$dB = 0$ $dE + \frac{\partial B}{\partial t} = 0$	$d * E = \frac{\rho}{\epsilon_0}$ $d * B - \frac{1}{c^2}\frac{\partial * E}{\partial t} = \mu_0 J$
Differential forms	**Potentials (any gauge)** **Any space (with topological restrictions) + time**	$B = dA$ $E = -d\phi - \frac{\partial A}{\partial t}$	$-d * (d\phi + \frac{\partial A}{\partial t}) = \frac{\rho}{\epsilon_0}$ $d * dA + \frac{1}{c^2}\frac{\partial}{\partial t} * (d\phi - \frac{\partial A}{\partial t}) = \mu_0 J$
	Potential (Lorenz Gauge) **Any space (with topological restrictions) + time** **spatial metric independent of time**	$B = dA$ $E = -d\phi - \frac{\partial A}{\partial t}$ $d * A = - * \frac{1}{c^2}\frac{\partial \phi}{\partial t}$	$*(-\Delta\phi + \frac{1}{c^2}\frac{\partial^2}{\partial t^2}\phi) = \frac{\rho}{\epsilon_0}$ $*(-\Delta A + \frac{1}{c^2}\frac{\partial^2 A}{\partial^2 t}) = \mu_0 J$

where

- In the vector formulation on Euclidean space + time, φ is the electrical potential, and **A** is the vector potential.

Relativistic Formulations

Formalism	Formulation	Homogeneous equations	Inhomogeneous equations
	Fields Minkowski space	$\partial_{[\alpha} F_{\beta\gamma]} = 0$	$\partial_\alpha F^{\alpha\beta} = \mu_0 J^\beta$
	Potentials (any gauge) **Minkowski space**	$F_{\alpha\beta} = 2\partial_{[\alpha} A_{\beta]}$	$2\partial_\alpha \partial^{[\alpha} A^{\beta]} = \mu_0 J^\beta$
	Potentials (Lorenz gauge) **Minkowski space**	$F_{\alpha\beta} = 2\partial_{[\alpha} A_{\beta]}$ $\partial_\alpha A^\alpha = 0$	$\partial_\alpha \partial^\alpha A^\beta = \mu_0 J^\beta$
Tensor calculus	**Fields** **Any space–time**	$\partial_{[\alpha} F_{\beta\gamma]} =$ $\nabla_{[\alpha} F_{\beta\gamma]} = 0$	$\frac{1}{\sqrt{-g}}\partial_\alpha(\sqrt{-g}F^{\alpha\beta}) =$ $\nabla_\alpha F^{\alpha\beta} = \mu_0 J^\beta$
	Potentials (any gauge) **Any space–time (with topological restrictions)**	$F_{\alpha\beta} = 2\partial_{[\alpha} A_{\beta]}$ $= 2\nabla_{[\alpha} A_{\beta]}$	$\frac{2}{\sqrt{-g}}\partial_\alpha(\sqrt{-g}g^{\alpha\mu}g^{\beta\nu}\partial_{[\mu}A_{\nu]}) =$ $2\nabla_\alpha(\nabla^{[\alpha}A^{\beta]}) = \mu_0 J^\beta$
	Potentials (Lorenz gauge) **Any space–time (with topological restrictions)**	$F_{\alpha\beta} = 2\partial_{[\alpha} A_{\beta]}$ $= 2\nabla_{[\alpha} A_{\beta]}$ $\nabla_\alpha A^\alpha = 0$	$\nabla_\alpha \nabla^\alpha A^\beta - R^\beta{}_\alpha A^\alpha = \mu_0 J^\beta$

Differential forms	**Fields** **Any space–time**	$\mathrm{d}F = 0$	$d \star F = \mu_0 J$
	Potentials (any gauge) **Any space–time (with topological restrictions)**	$F = \mathrm{d}A$	$d \star dA = \mu_0 J$
	Potentials (Lorenz gauge) **Any space–time (with topological restrictions)**	$F = dA$ $d \star A = 0$	$\star \Box A = \mu_0 J$

The Maxwell equations can also be formulated on a space-time like Minkowski space where space and time are treated on equal footing. The direct space–time formulations make manifest that the Maxwell equations are relativistically invariant. Because of this symmetry electric and magnetic field are treated on equal footing and are recognised as components of the Faraday tensor. This reduces the four Maxwell equations to two, which simplifies the equations, although we can no longer use the familiar vector formulation. In fact the Maxwell equations in the space + time formulation are not Galileo invariant and have Lorenz invariance a hidden symmetry. This was a major source of inspiration for the development of relativity theory. The space + time formulation is not a non-relativistic approximation, however, they describe the same physics by simply renaming variables. For this reason the relativistic invariant equations are usually simply called the Maxwell equations as well.

- In the tensor calculus formulation, the electromagnetic tensor $F_{\alpha\beta}$ is an antisymmetric covariant rank 2 tensor; the four-potential, A_α, is a covariant vector; the current, J^α, is a vector; the square brackets, [ ], denote antisymmetrization of indices; ∂_α is the derivative with respect to the coordinate, x^α. In Minkowski space coordinates are chosen with respect to an inertial frame; $(x^\alpha) = (ct,x,y,z)$, so that the metric tensor used to raise and lower indices is $\eta_{\alpha\beta} = \mathrm{diag}(1,-1,-1,-1)$. The d'Alembert operator on Minkowski space is $\Box = \partial_\alpha \partial^\alpha$ as in the vector formulation. In general spacetimes, the coordinate system x^α is arbitrary, the covariant derivative ∇_α, the Ricci tensor, $R_{\alpha\beta}$ and raising and lowering of indices are defined by the Lorentzian metric, $g_{\alpha\beta}$ and the d'Alembert operator is defined as $\Box = \nabla_\alpha \nabla^\alpha$. The topological restriction is that the second real cohomology group of the space vanishes. Note that this is violated for Minkowski space with a line removed, which can model a (flat) space-time with a point-like monopole on the complement of the line.

- In the differential form formulation on arbitrary space times, $F = F_{\alpha\beta} \mathrm{d}x^\alpha \wedge \mathrm{d}x^\beta$ is the electromagnetic tensor considered as a 2-form, $A = A_\alpha \mathrm{d}x^\alpha$ is the potential 1-form, J is the current 3-form, d is the exterior derivative, and $\star$ is the Hodge star on forms defined by the Lorentzian metric of space–time. Note that in the special case of 2-forms such as F, the Hodge star $\star$ only depends on the metric up to a local scale. This means that, as formulated, the differential form field equations are conformally invariant, but the Lorenz gauge condition breaks conformal invariance. The operator $\Box = (-\star \mathrm{d} \star \mathrm{d} - \mathrm{d} \star \mathrm{d} \star)$ is the d'Alembert–Laplace–Beltrami operator on 1-forms on an arbitrary Lorentzian space–time. The topological condition is again that the second real cohomology group is trivial. By the isomorphism with the second de Rham cohomology this condition means that every closed 2 form is exact.

Other formalisms include the geometric algebra formulation and a matrix representation of Maxwell's equations. Historically, a quaternionic formulation was used.

Solutions

Maxwell's equations are partial differential equations that relate the electric and magnetic fields to each other and to the electric charges and currents. Often, the charges and currents are themselves dependent on the electric and magnetic fields via the Lorentz force equation and the constitutive relations. These all form a set of coupled partial differential equations, which are often very difficult to solve. In fact, the solutions of these equations encompass all the diverse phenomena in the entire field of classical electromagnetism. A thorough discussion is far beyond the scope of the article, but some general notes follow.

Like any differential equation, boundary conditions and initial conditions are necessary for a unique solution. For example, even with no charges and no currents anywhere in spacetime, many solutions to Maxwell's equations are possible, not just the obvious solution $E = B = 0$. Another solution is E = constant, B = constant, while yet other solutions have electromagnetic waves filling spacetime. In some cases, Maxwell's equations are solved through infinite space, and boundary conditions are given as asymptotic limits at infinity. In other cases, Maxwell's equations are solved in just a finite region of space, with appropriate boundary conditions on that region: For example, the boundary could be an artificial absorbing boundary representing the rest of the universe, or periodic boundary conditions, or (as with a waveguide or cavity resonator) the boundary conditions may describe the walls that isolate a small region from the outside world.

Jefimenko's equations (or the closely related Liénard–Wiechert potentials) are the explicit solution to Maxwell's equations for the electric and magnetic fields created by any given distribution of charges and currents. It assumes specific initial conditions to obtain the so-called "retarded solution", where the only fields present are the ones created by the charges. Jefimenko's equations are not so helpful in situations when the charges and currents are themselves affected by the fields they create.

Numerical methods for differential equations can be used to approximately solve Maxwell's equations when an exact solution is impossible. These methods usually require a computer, and include the finite element method and finite-difference time-domain method.

Maxwell's equations *seem* overdetermined, in that they involve six unknowns (the three components of E and B) but eight equations (one for each of the two Gauss's laws, three vector components each for Faraday's and Ampere's laws). (The currents and charges are not unknowns, being freely specifiable subject to charge conservation.) This is related to a certain limited kind of redundancy in Maxwell's equations: It can be proven that any system satisfying Faraday's law and Ampere's law *automatically* also satisfies the two Gauss's laws, as long as the system's initial condition does. This explanation was first introduced by Julius Adams Stratton in 1941. Although it is possible to simply ignore the two Gauss's laws in a numerical algorithm (apart from the initial conditions), the imperfect precision of the calculations can lead to ever-increasing violations of those laws. By introducing dummy variables characterizing these violations, the four equations become not overdetermined after all. The resulting formulation can lead to more accurate algorithms that take all four laws into account.

Limitations of the Maxwell Equations as a Theory of Electromagnetism

While Maxwell's equations (along with the rest of classical electromagnetism) are extraordinarily successful at explaining and predicting a variety of phenomena, they are not exact, but approximations. In some special situations, they can be noticeably inaccurate. Examples include extremely strong fields and extremely short distances. Moreover, various phenomena occur in the world even though Maxwell's equations predict them to be impossible, such as "nonclassical light" and quantum entanglement of electromagnetic fields. Finally, any phenomenon involving individual photons, such as the photoelectric effect, Planck's law, the Duane–Hunt law, single-photon light detectors, etc., would be difficult or impossible to explain if Maxwell's equations were exactly true, as Maxwell's equations do not involve photons. For the most accurate predictions in all situations, Maxwell's equations have been superseded by quantum electrodynamics.

Variations

Popular variations on the Maxwell equations as a classical theory of electromagnetic fields are relatively scarce because the standard equations have stood the test of time remarkably well.

Magnetic Monopoles

Maxwell's equations posit that there is electric charge, but no magnetic charge (also called magnetic monopoles), in the universe. Indeed, magnetic charge has never been observed (despite extensive searches) and may not exist. If they did exist, both Gauss's law for magnetism and Faraday's law would need to be modified, and the resulting four equations would be fully symmetric under the interchange of electric and magnetic fields.

Maxwell's Equations are

$$\nabla . E = \frac{\rho}{\varepsilon_0} \qquad (Gauss's\ Law)$$

$$\nabla . B = 0$$

$$\nabla \times E = -\frac{\partial B}{\partial t} \quad (Faraday's\ Law)$$

$$\nabla \times B = \mu_0 J + \mu_0 \varepsilon_0 \frac{\partial E}{\partial t} \quad (Ampere's\ Law)$$

where, E & B are the electric and magnetic field components;

ρ is the charge density;

μ_0 & ε_0 are the permeability and permittivity of the free space respectively;

and $\mu_0 \varepsilon_0 = \frac{1}{c^2}$ In free space, i.e. no charge and current, Maxwell's equations become,

$$\nabla . E = 0 \quad ; \quad \nabla \times E = -\frac{\partial}{\partial}$$

$$\nabla . B = 0 \quad ; \quad \nabla \times B = \frac{}{c} \frac{\partial}{\partial t}$$

Decoupling the electric and magnetic field from the above equations

$$\nabla^2 E = \frac{1}{c^2}\frac{\partial^2 E}{\partial t^2} \quad ; \quad \nabla^2 B = \frac{1}{c^2}\frac{\partial^2 B}{\partial t^2}$$

Solution of these are,

$$\bar{E}(x,t) = E_0 e^{i(Kx-\omega t)}\hat{j}$$

$$\bar{B}(x,t) = \frac{1}{c}E_0 e^{i(Kx-\omega t)}\hat{k}$$

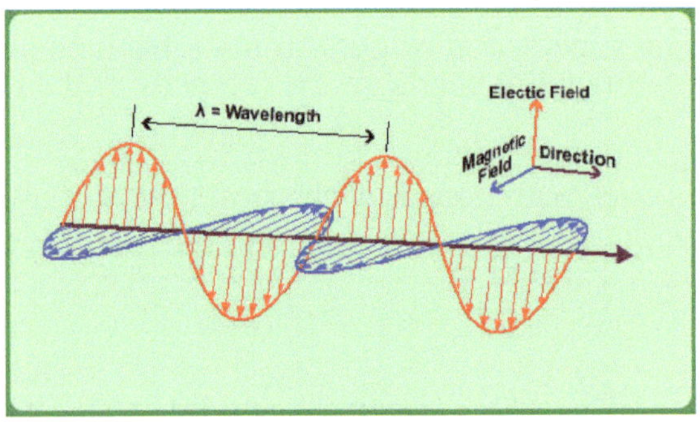

Figure (a)

where $K = \frac{2\pi}{\lambda}$ is the propagation constant.

In the beginning of the 20th century, several experiments and observations could not be explained by Classical Physics. These are:

- Photoelectric effect
- Compton effect
- Phenomena on atomic scales such as classical explanation of Rutherford model, discrete energy levels of atoms proposed by Bohr.

These facts forced the scientist to search for a near explanation of the mechanics.

- Photoelectric effect

In 1886 and 1887, Heinrich Hertz showed that an electric discharge between the two electrodes shown in the Figure –b) occurs when ultraviolet light falls on one of the electrodes.

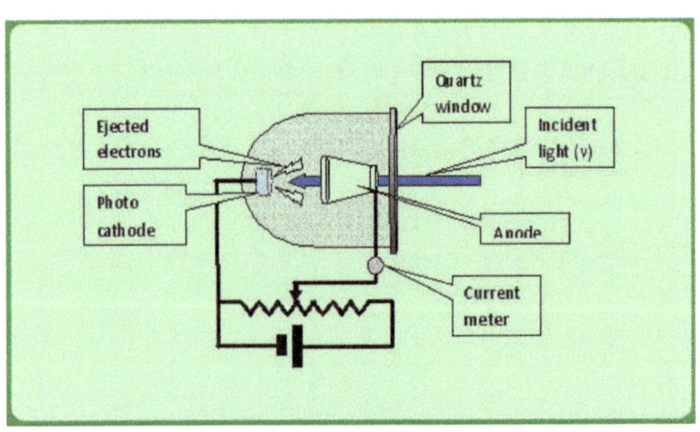

Figure (b)

In this a metallic cathode and anode are placed in a vacuum chamber having a quartz (clear to UV light) window. UV light having with frequency (n) is incident on cathode and the current generated by this light is measured through a current meter.

The first of these phenomena was discovered by Hertz and Hallwachs in 1887. The second was announced first by Philipp Lenard in 1900.

Observations

1. When the potential across anode with respect to cathode is varied, it is observed that after a certain potential, the photocurrent starts and increased and slowly gets saturated as shown in Figure – c.
2. The magnitude of the current depends on the intensity of the light.
3. Stopping potential depends on the frequency of the light. (K_{max} = maximum kinetic energy of the photoelectron $= eV_0$).

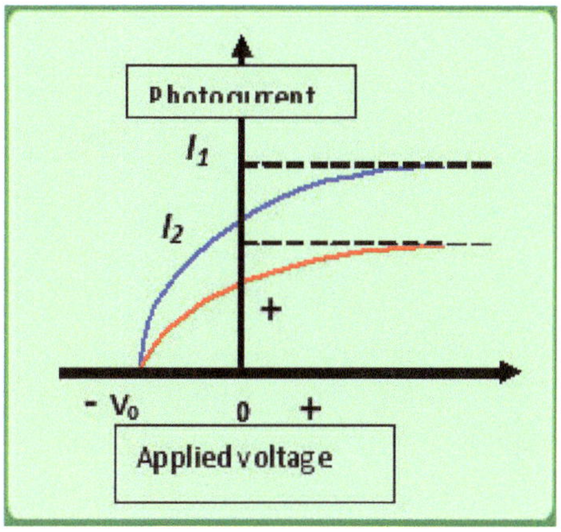

Figure (c)

These observations cannot be explained by classical theory. The reasons are:

- In the wave picture, the amplitude of the electric field E_0 (Figure -a) increases as the light intensity is increased. The force exerted by this electric field on the electron is $e\vec{E}$. It means the kinetic energy of the ejected photoelectron should increase if the intensity is increased. However, from the figure, it is clear that $K_{max} = eV_0$ which is independent of light intensity.
- According to the wave theory, photoelectron should exit at any frequency of light if the exerted force is greater than the binding force. However, observation is such that the photoelectrons for a particular metal (cathode) are ejected only when the light frequency is greater than a particular value, a cut off frequency ν_0.

To overcome these discrepancies, Einstein took the idea given by Planck and assumed, that

- The light energy is a bundle of energy localized in a small volume of space.
- It remains localized even it moves with a velocity of light c. He termed the content of the energy E as "PHOTON", and is related to the frequency ν as given in Equation – 5.

$E = hv$

- For the photoelectric effect, he assumed that one of such photons is completely absorbed by one electron of the photocathode as if particle-particle interaction and derived the kinetic energy of the electron as given in Equation below.

$K = hv - W_0$

where, W_0 is the work functions of the photocathode.

The particle nature of light (quantization of radiation) was again realized to explain the Compton's experiment in 1923.

The X-ray of wavelength λ_0 was incident on a target. With a crystal diffractometer, the wavelengths of the scattered X-ray were measured by changing the angle θ.

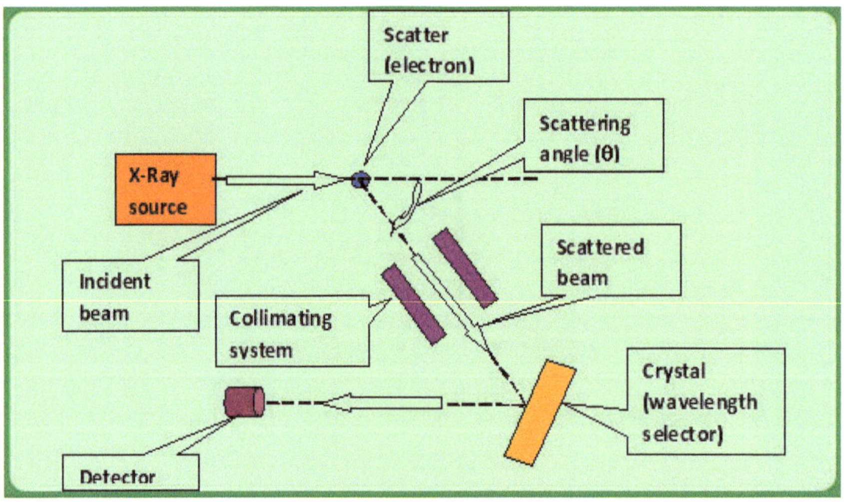

Figure (d)

- The scattered radiation consists of the primary wavelength λ_0 and a shifted wavelength.
- The Compton shift $\Delta\lambda = \lambda - \lambda_0$ varies with the scattering angle.
- The classical theory predicts that if the radiation of wavelength l incidents on a particle, it scatters and emits radiation of the same wavelength l. No other wavelength is predicted by this theory.
- Using the particle nature of the radiation, from the momentum and energy conservation, the Compton equation is derived to explain the observation as in Equation below,

$$\Delta\lambda = \frac{h}{m_0 c}(1 - \cos\theta) = \lambda_c(1 - \cos\theta)$$

where, $\lambda_c = \frac{h}{m_0 c} = 2.43 \times 10^{-12} meter = 0.0243 \mathring{A}$

Electromagnetic Radiation

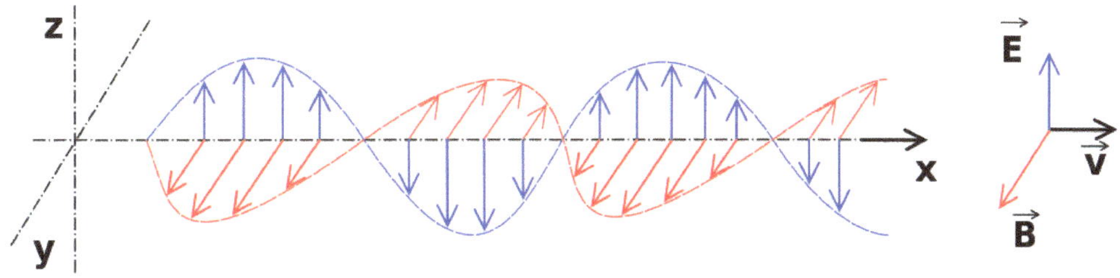

The electromagnetic waves that compose electromagnetic radiation can be imagined as a self-propagating transverse oscillating wave of electric and magnetic fields. This diagram shows a plane linearly polarized EMR wave propagating from left to right (X axis). The electric field is in a vertical plane (Z axis) and the magnetic field in a horizontal plane (Y axis). The electric and magnetic fields in EMR waves are always in phase and at 90 degrees to each other.

In physics, electromagnetic radiation (EM radiation or EMR) refers to the waves (or their quanta, photons) of the electromagnetic field, propagating (radiating) through space carrying electromagnetic radiant energy. It includes radio waves, microwaves, infrared, (visible) light, ultraviolet, X-, and gamma radiation.

Classically, electromagnetic radiation consists of electromagnetic waves, which are synchronized oscillations of electric and magnetic fields that propagate at the speed of light through a vacuum. The oscillations of the two fields are perpendicular to each other and perpendicular to the direction of energy and wave propagation, forming a transverse wave. The wavefront of electromagnetic waves emitted from a point source (such as a lightbulb) is a sphere. The position of an electromagnetic wave within the electromagnetic spectrum can be characterized by either its frequency of oscillation or its wavelength. The electromagnetic spectrum includes, in order of increasing frequency and decreasing wavelength: radio waves, microwaves, infrared radiation, visible light, ultraviolet radiation, X-rays and gamma rays.

Electromagnetic waves are produced whenever charged particles are accelerated, and these waves can subsequently interact with other charged particles. EM waves carry energy, momentum and angular momentum away from their source particle and can impart those quantities to matter with which they interact. Quanta of EM waves are called photons, whose rest mass is zero, but whose energy, or equivalent total (relativistic) mass, is not zero so they are still affected by gravity. Electromagnetic radiation is associated with those EM waves that are free to propagate themselves ("radiate") without the continuing influence of the moving charges that produced them, because they have achieved sufficient distance from those charges. Thus, EMR is sometimes referred to as the far field. In this language, the near fiel*d* refers to EM fields near the charges and current that directly produced them, specifically, electromagnetic induction and electrostatic induction phenomena.

In the quantum theory of electromagnetism, EMR consists of photons, the elementary particles responsible for all electromagnetic interactions. Quantum effects provide additional sources of EMR, such as the transition of electrons to lower energy levels in an atom and black-body radiation. The energy of an individual photon is quantized and is greater for photons of higher frequency. This relationship is given by Planck's equation $E = hv$, where E is the energy per photon, v is the frequency of the photon, and h is Planck's constant. A single gamma ray photon, for example, might carry ~100,000 times the energy of a single photon of visible light.

The effects of EMR upon chemical compounds and biological organisms depend both upon the radiation's power and its frequency. EMR of visible or lower frequencies (i.e., visible light, infrared, microwaves, and radio waves) is called *non-ionizing radiation*, because its photons do not individually have enough energy to ionize atoms or molecules. The effects of these radiations on chemical systems and living tissue are caused primarily by heating effects from the combined energy transfer of many photons. In contrast, high ultraviolet, X-rays and gamma rays are called *ionizing radiation* since individual photons of such high frequency have enough energy to ionize molecules or break chemical bonds. These radiations have the ability to cause chemical reactions and damage living cells beyond that resulting from simple heating, and can be a health hazard.

Physics

Theory

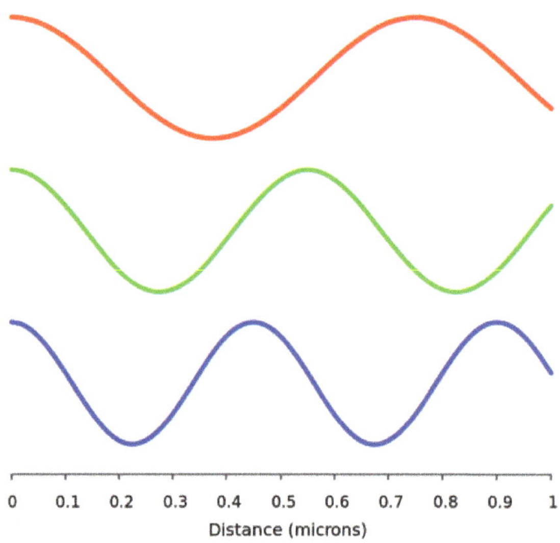

Shows the relative wavelengths of the electromagnetic waves of three different colours of light (blue, green, and red) with a distance scale in micrometers along the x-axis

Maxwell's Equations

Maxwell derived a wave form of the electric and magnetic equations, thus uncovering the wave-like nature of electric and magnetic fields and their symmetry. Because the speed of EM waves predicted by the wave equation coincided with the measured speed of light, Maxwell concluded that light itself is an EM wave. Maxwell's equations were confirmed by Heinrich Hertz through experiments with radio waves.

According to Maxwell's equations, a spatially varying electric field is always associated with a magnetic field that changes over time. Likewise, a spatially varying magnetic field is associated with specific changes over time in the electric field. In an electromagnetic wave, the changes in the electric field are always accompanied by a wave in the magnetic field in one direction, and vice versa. This relationship between the two occurs without either type field causing the other; rather, they occur together in the same way that time and space changes occur together and are interlinked in special relativity. In fact, magnetic fields may be viewed as relativistic distortions of electric fields, so the close relationship between space and time changes here is more than an analogy. Together, these fields form a propagating electromagnetic wave, which moves out into space and need never again affect the source. The distant EM field formed in this way by the acceleration of a charge carries energy with it that "radiates" away through space, hence the term.

Near and Far Fields

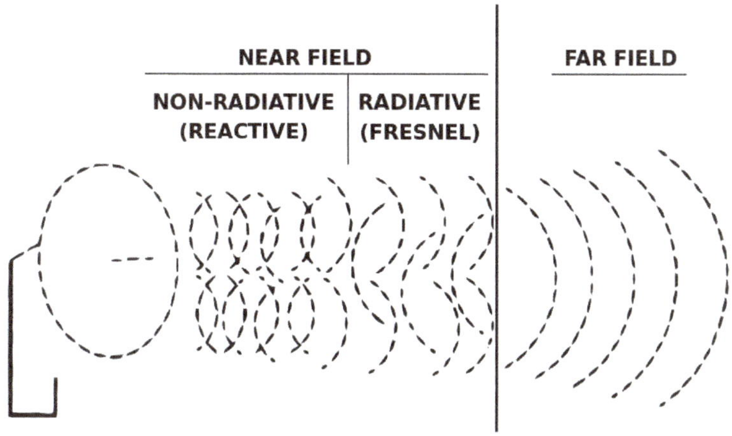

In electromagnetic radiation (such as microwaves from an antenna, shown here) the term applies only to the parts of the electromagnetic field that radiate into infinite space and decrease in intensity by an inverse-square law of power, so that the total radiation energy that crosses through an imaginary spherical surface is the same, no matter how far away from the antenna the spherical surface is drawn. Electromagnetic radiation thus includes the far field part of the electromagnetic field around a transmitter. A part of the "near-field" close to the transmitter, forms part of the changing electromagnetic field, but does not count as electromagnetic radiation.

Maxwell's equations established that some charges and currents ("sources") produce a local type of electromagnetic field near them that does *not* have the behaviour of EMR. Currents directly produce a magnetic field, but it is of a magnetic dipole type that dies out with distance from the current. In a similar manner, moving charges pushed apart in a conductor by a changing electrical potential (such as in an antenna) produce an electric dipole type electrical field, but this also declines with distance. These fields make up the near-field near the EMR source. Neither of these behaviours are responsible for EM radiation. Instead, they cause electromagnetic field behaviour that only efficiently transfers power to a receiver very close to the source, such as the magnetic induction inside a transformer, or the feedback behaviour that happens close to the coil of a metal detector. Typically, near-fields have a powerful effect on their own sources, causing an increased "load" (decreased electrical reactance) in the source or transmitter, whenever energy is withdrawn from the EM field by a receiver. Otherwise, these fields do not "propagate" freely out into space, carrying their energy away without distance-limit, but rather oscillate, returning their energy to the transmitter if it is not received by a receiver.

By contrast, the EM far-field is composed of *radiation* that is free of the transmitter in the sense that (unlike the case in an electrical transformer) the transmitter requires the same power to send these changes in the fields out, whether the signal is immediately picked up or not. This distant part of the electromagnetic field *is* "electromagnetic radiation" (also called the far-field). The far-fields propagate (radiate) without allowing the transmitter to affect them. This causes them to be independent in the sense that their existence and their energy, after they have left the transmitter, is completely independent of both transmitter and receiver. Because such waves conserve the amount of energy they transmit through any spherical boundary surface drawn around their source, and because such surfaces have an area that is defined by the square of the distance from the source, the power of EM radiation always varies according to an inverse-square law. This is in contrast to dipole parts of the EM field close to the source (the near-field), which varies in power according to an inverse cube power law, and thus does *not* transport a conserved amount of energy over distances, but instead fades with distance, with its energy (as noted) rapidly returning to the transmitter or absorbed by a nearby receiver (such as a transformer secondary coil).

The far-field (EMR) depends on a different mechanism for its production than the near-field, and upon different terms in Maxwell's equations. Whereas the magnetic part of the near-field is due to currents in the source, the magnetic field in EMR is due only to the local change in the electric field. In a similar way, while the electric field in the near-field is due directly to the charges and charge-separation in the source, the electric field in EMR is due to a change in the local magnetic field. Both processes for producing electric and magnetic EMR fields have a different dependence on distance than do near-field dipole electric and magnetic fields. That is why the EMR type of EM field becomes dominant in power "far" from sources. The term "far from sources" refers to how far from the source (moving at the speed of light) any portion of the outward-moving EM field is located, by the time that source currents are changed by the varying source potential, and the source has therefore begun to generate an outwardly moving EM field of a different phase.

A more compact view of EMR is that the far-field that composes EMR is generally that part of the EM field that has traveled sufficient distance from the source, that it has become completely disconnected from any feedback to the charges and currents that were originally responsible for it. Now independent of the source charges, the EM field, as it moves farther away, is dependent only upon the accelerations of the charges that produced it. It no longer has a strong connection to the direct fields of the charges, or to the velocity of the charges (currents).

In the Liénard–Wiechert potential formulation of the electric and magnetic fields due to motion of a single particle (according to Maxwell's equations), the terms associated with acceleration of the particle are those that are responsible for the part of the field that is regarded as electromagnetic radiation. By contrast, the term associated with the changing static electric field of the particle and the magnetic term that results from the particle's uniform velocity, are both associated with the electromagnetic near-field, and do not comprise EM radiation.

Properties

Electrodynamics is the physics of electromagnetic radiation, and electromagnetism is the physical phenomenon associated with the theory of electrodynamics. Electric and magnetic fields obey the properties of superposition. Thus, a field due to any particular particle or time-varying electric or magnetic field contributes to the fields present in the same space due to other causes. Further, as they are vector fields, all magnetic and electric field vectors add together according to vector addition. For example, in optics two or more coherent lightwaves may interact and by constructive or destructive interference yield a resultant irradiance deviating from the sum of the component irradiances of the individual lightwaves.

Since light is an oscillation it is not affected by traveling through static electric or magnetic fields in a linear medium such as a vacuum. However, in nonlinear media, such as some crystals, interactions can occur between light and static electric and magnetic fields — these interactions include the Faraday effect and the Kerr effect.

In refraction, a wave crossing from one medium to another of different density alters its speed and direction upon entering the new medium. The ratio of the refractive indices of the media determines the degree of refraction, and is summarized by Snell's law. Light of composite wavelengths (natural sunlight) disperses into a visible spectrum passing through a prism, because of the wavelength-dependent refractive index of the prism material (dispersion); that is, each component wave within the composite light is bent a different amount.

EM radiation exhibits both wave properties and particle properties at the same time. Both wave and particle characteristics have been confirmed in many experiments. Wave characteristics are more apparent when EM radiation is measured over relatively large timescales and over large distances while particle characteristics are more evident when measuring small timescales and distances. For example, when electromagnetic radiation is absorbed by matter, particle-like properties will be more obvious when the average number of photons in the cube of the relevant wavelength is much smaller than 1. It is not too difficult to experimentally observe non-uniform deposition of energy when light is absorbed, however this alone is not evidence of "particulate" behavior. Rather, it reflects the quantum nature of *matter*. Demonstrating that the light itself is quantized, not merely its interaction with matter, is a more subtle affair.

Some experiments display both the wave and particle natures of electromagnetic waves, such as the self-interference of a single photon. When a single photon is sent through an interferometer, it passes through both paths, interfering with itself, as waves do, yet is detected by a photomultiplier or other sensitive detector only once.

A quantum theory of the interaction between electromagnetic radiation and matter such as electrons is described by the theory of quantum electrodynamics.

Electromagnetic waves can be polarized, reflected, refracted, diffracted or interfere with each other.

Wave Model

Electromagnetic radiation is a transverse wave, meaning that its oscillations are perpendicular to the direction of energy transfer and travel. The electric and magnetic parts of the field stand in a fixed ratio of strengths in order to satisfy the two Maxwell equations that specify how one is produced from the other. These E and B fields are also in phase, with both reaching maxima and minima at the same points in space. A common misconception is that the E and B fields in electromagnetic radiation are out of phase because a change in one produces the other, and this would produce a phase difference between them as sinusoidal functions (as indeed happens in electromagnetic induction, and in the near-field close to antennas). However, in the far-field EM radiation which is described by the two source-free Maxwell curl operator equations, a more correct description is that a time-change in one type of field is proportional to a space-change in the other. These derivatives require that the E and B fields in EMR are in-phase.

An important aspect of light's nature is its frequency. The frequency of a wave is its rate of oscillation and is measured in hertz, the SI unit of frequency, where one hertz is equal to one oscillation per second. Light usually has multiple frequencies that sum to form the resultant wave. Different frequencies undergo different angles of refraction, a phenomenon known as dispersion.

A wave consists of successive troughs and crests, and the distance between two adjacent crests or troughs is called the wavelength. Waves of the electromagnetic spectrum vary in size, from very long radio waves the size of buildings to very short gamma rays smaller than atom nuclei. Frequency is inversely proportional to wavelength, according to the equation:

$$v = f\lambda$$

where v is the speed of the wave (c in a vacuum, or less in other media), f is the frequency and λ is the wavelength. As waves cross boundaries between different media, their speeds change but their frequencies remain constant.

Electromagnetic waves in free space must be solutions of Maxwell's electromagnetic wave equation. Two main classes of solutions are known, namely plane waves and spherical waves. The plane waves may be viewed as the limiting case of spherical waves at a very large (ideally infinite) distance from the source. Both types of waves can have a waveform which is an arbitrary time function (so long as it is sufficiently differentiable to conform to the wave equation). As with any time function, this can be decomposed by means of Fourier analysis into its frequency spectrum, or individual sinusoidal components, each of which contains a single frequency, amplitude and phase. Such a component wave is said to be *monochromatic*. A monochromatic electromagnetic wave can be characterized by its frequency or wavelength, its peak amplitude, its phase relative to some reference phase, its direction of propagation and its polarization.

Interference is the superposition of two or more waves resulting in a new wave pattern. If the fields have components in the same direction, they constructively interfere, while opposite directions cause destructive interference. An example of interference caused by EMR is electromagnetic interference (EMI) or as it is more commonly known as, radio-frequency interference (RFI). Additionally, multiple polarization signals can be combined (i.e. interfered) to form new states of polarization, which is known as parallel polarization state generation.

The energy in electromagnetic waves is sometimes called radiant energy.

Particle Model and Quantum Theory

An anomaly arose in the late 19th century involving a contradiction between the wave theory of light and measurements of the electromagnetic spectra that were being emitted by thermal radiators known as black bodies. Physicists struggled with this problem, which later became known as the ultraviolet catastrophe, unsuccessfully for many years. In 1900, Max Planck developed a new theory of black-body radiation that explained the observed spectrum. Planck's theory was based on the idea that black bodies emit light (and other electromagnetic radiation) only as discrete bundles or packets of energy. These packets were called quanta. Later, Albert Einstein proposed that light quanta be regarded as real particles. Later the particle of light was given the name photon, to correspond with other particles being described around this time, such as the electron and proton. A photon has an energy, E, proportional to its frequency, f, by

$$E = hf = \frac{hc}{\lambda}$$

where h is Planck's constant, λ is the wavelength and c is the speed of light. This is sometimes known as the Planck–Einstein equation. In quantum theory the energy of the photons is thus directly proportional to the frequency of the EMR wave.

Likewise, the momentum p of a photon is also proportional to its frequency and inversely proportional to its wavelength:

$$p = \frac{E}{c} = \frac{hf}{c} = \frac{h}{\lambda}.$$

The source of Einstein's proposal that light was composed of particles (or could act as particles in some circumstances) was an experimental anomaly not explained by the wave theory: the photoelectric effect, in which light striking a metal surface ejected electrons from the surface, causing an electric current to flow across an applied voltage. Experimental measurements demonstrated that the energy of individual ejected electrons was proportional to the *frequency*, rather than the *intensity*, of the light. Furthermore, below a certain minimum frequency, which depended on the particular metal, no current would flow regardless of the intensity. These observations appeared to contradict the wave theory, and for years physicists tried in vain to find an explanation. In 1905, Einstein explained this puzzle by resurrecting the particle theory of light to explain the observed effect. Because of the preponderance of evidence in favor of the wave theory, however, Einstein's ideas were met initially with great skepticism among established physicists. Eventually Einstein's explanation was accepted as new particle-like behavior of light was observed, such as the Compton effect.

As a photon is absorbed by an atom, it excites the atom, elevating an electron to a higher energy level (one that is on average farther from the nucleus). When an electron in an excited molecule or atom descends to a lower energy level, it emits a photon of light at a frequency corresponding to the energy difference. Since the energy levels of electrons in atoms are discrete, each element and each molecule emits and absorbs its own characteristic frequencies. Immediate photon emission is called fluorescence, a type of photoluminescence. An example is visible light emitted from fluorescent paints, in response to ultraviolet (blacklight). Many other fluorescent emissions are known in spectral bands other than visible light. Delayed emission is called phosphorescence.

Wave–particle Duality

The modern theory that explains the nature of light includes the notion of wave–particle duality. More generally, the theory states that everything has both a particle nature and a wave nature, and various experiments can be done to bring out one or the other. The particle nature is more easily discerned using an object with a large mass. A bold proposition by Louis de Broglie in 1924 led the scientific community to realize that electrons also exhibited wave–particle duality.

Wave and Particle Effects of Electromagnetic Radiation

Together, wave and particle effects fully explain the emission and absorption spectra of EM radiation. The matter-composition of the medium through which the light travels determines the nature of the absorption and emission spectrum. These bands correspond to the allowed energy levels in the atoms. Dark bands in the absorption spectrum are due to the atoms in an intervening medium between source and observer. The atoms absorb certain frequencies of the light between emitter and detector/eye, then emit them in all directions. A dark band appears to the detector, due to the radiation scattered out of the beam. For instance, dark bands in the light emitted by a distant star are due to the atoms in the star's atmosphere. A similar phenomenon occurs for emission, which is seen when an emitting gas glows due to excitation of the atoms from any mechanism, including heat. As electrons descend to lower energy levels, a spectrum is emitted that represents the jumps between the energy levels of the electrons, but lines are seen because again emission happens only at particular energies after excitation. An example is the emission spectrum of nebulae. Rapidly moving electrons are most sharply accelerated when they encounter a region of force, so they are responsible for producing much of the highest frequency electromagnetic radiation observed in nature.

These phenomena can aid various chemical determinations for the composition of gases lit from behind (absorption spectra) and for glowing gases (emission spectra). Spectroscopy (for example) determines what chemical elements comprise a particular star. Spectroscopy is also used in the determination of the distance of a star, using the red shift.

Propagation Speed

Any electric charge that accelerates, or any changing magnetic field, produces electromagnetic radiation. Electromagnetic information about the charge travels at the speed of light. Accurate treatment thus incorporates a concept known as retarded time, which adds to the expressions for the electrodynamic electric field and magnetic field. These extra terms are responsible for electromagnetic radiation.

When any wire (or other conducting object such as an antenna) conducts alternating current, electromagnetic radiation is propagated at the same frequency as the current. In many such situations it is possible to identify an electrical dipole moment that arises from separation of charges due to the exciting electrical potential, and this dipole moment oscillates in time, as the charges move back and forth. This oscillation at a given frequency gives rise to changing electric and magnetic fields, which then set the electromagnetic radiation in motion.

At the quantum level, electromagnetic radiation is produced when the wavepacket of a charged particle oscillates or otherwise accelerates. Charged particles in a stationary state do not move, but a superposition of such states may result in a transition state that has an electric dipole moment that oscillates in time. This oscillating dipole moment is responsible for the phenomenon of radiative transition between quantum states of a charged particle. Such states occur (for example) in atoms when photons are radiated as the atom shifts from one stationary state to another.

As a wave, light is characterized by a velocity (the speed of light), wavelength, and frequency. As particles, light is a stream of photons. Each has an energy related to the frequency of the wave given by Planck's relation $E = hf$, where E is the energy of the photon, $h = 6.626 \times 10^{-34}$ J·s is Planck's constant, and f is the frequency of the wave.

One rule is obeyed regardless of circumstances: EM radiation in a vacuum travels at the speed of light, *relative to the observer*, regardless of the observer's velocity. (This observation led to Einstein's development of the theory of special relativity.)

In a medium (other than vacuum), velocity factor or refractive index are considered, depending on frequency and application. Both of these are ratios of the speed in a medium to speed in a vacuum.

Special Theory of Relativity

By the late nineteenth century, various experimental anomalies could not be explained by the simple wave theory. One of these anomalies involved a controversy over the speed of light. The speed of light and other EMR predicted by Maxwell's equations did not appear unless the equations were modified in a way first suggested by FitzGerald and Lorentz, or else otherwise that speed would depend on the speed of observer relative to the "medium" (called luminiferous aether) which supposedly "carried" the electromagnetic wave (in a manner analogous to the way air carries sound waves). Experiments failed to find any observer effect. In 1905, Einstein proposed that space and time appeared to be velocity-changeable entities for light propagation and all other processes and laws. These changes accounted for the constancy of the speed of light and all electromagnetic radiation, from the viewpoints of all observers—even those in relative motion.

History of Discovery

Electromagnetic radiation of wavelengths other than those of visible light were discovered in the early 19th century. The discovery of infrared radiation is ascribed to astronomer William Herschel, who published his results in 1800 before the Royal Society of London. Herschel used a glass prism to refract light from the Sun and detected invisible rays that caused heating beyond the red part of the spectrum, through an increase in the temperature recorded with a thermometer. These "calorific rays" were later termed infrared.

In 1801, German physicist Johann Wilhelm Ritter discovered ultraviolet in an experiment similar to Hershel's, using sunlight and a glass prism. Ritter noted that invisible rays near the violet edge of a solar spectrum dispersed by a triangular prism darkened silver chloride preparations more quickly than did the nearby violet light. Ritter's experiments were an early precursor to what would become photography. Ritter noted that the ultraviolet rays (which at first were called "chemical rays") were capable of causing chemical reactions.

In 1862-4 James Clerk Maxwell developed equations for the electromagnetic field which suggested that waves in the field would travel with a speed that was very close to the known speed of light. Maxwell therefore suggested that visible light (as well as invisible infrared and ultraviolet rays by inference) all consisted of propagating disturbances (or radiation) in the electromagnetic field. Radio waves were first produced deliberately by Heinrich Hertz in 1887, using electrical circuits calculated to produce oscillations at a much lower frequency than that of visible light, following recipes for producing oscillating charges and currents suggested by Maxwell's equations. Hertz also developed ways to detect these waves, and produced and characterized what were later termed radio waves and microwaves.

Wilhelm Röntgen discovered and named X-rays. After experimenting with high voltages applied to an evacuated tube on 8 November 1895, he noticed a fluorescence on a nearby plate of coated glass. In one month, he discovered X-rays' main properties.

The last portion of the EM spectrum to be discovered was associated with radioactivity. Henri Becquerel found that uranium salts caused fogging of an unexposed photographic plate through a covering paper in a manner similar to X-rays, and Marie Curie discovered that only certain elements gave off these rays of energy, soon discovering the intense radiation of radium. The radiation from pitchblende was differentiated into alpha rays (alpha particles) and beta rays (beta particles) by Ernest Rutherford through simple experimentation in 1899, but these proved to be charged particulate types of radiation. However, in 1900 the French scientist Paul Villard discovered a third neutrally charged and especially penetrating type of radiation from radium, and after he described it, Rutherford realized it must be yet a third type of radiation, which in 1903 Rutherford named gamma rays. In 1910 British physicist William Henry Bragg demonstrated that gamma rays are electromagnetic radiation, not particles, and in 1914 Rutherford and Edward Andrade measured their wavelengths, finding that they were similar to X-rays but with shorter wavelengths and higher frequency, although a 'cross-over' between X and gamma rays makes it possible to have X-rays with a higher energy (and hence shorter wavelength) than gamma rays and vice versa. The origin of the ray differentiates them, gamma rays tend to be a natural phenomena originating from the unstable nucleus of an atom and X-rays are electrically generated (and hence man-made) unless they are as a result of bremsstrahlung X-radiation caused by the interaction of fast moving particles (such as beta particles) colliding with certain materials, usually of higher atomic numbers.

Electromagnetic Spectrum

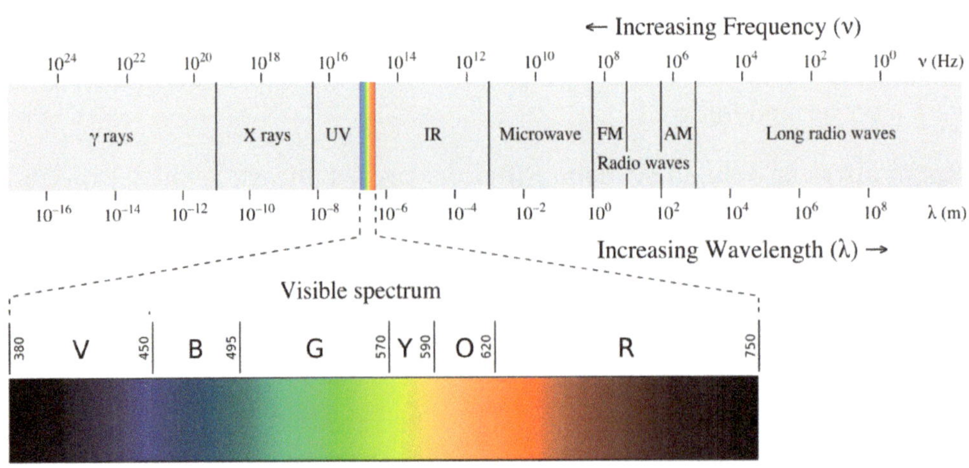

Electromagnetic spectrum with visible light highlighted

CLASS	FREQUENCY	WAVELENGTH	ENERGY
γ	300 EHz	1 pm	1.24 MeV
HX	30 EHz	10 pm	124 keV
	3 EHz	100 pm	12.4 keV
SX	300 PHz	1 nm	1.24 keV
EUV	30 PHz	10 nm	124 eV
NUV	3 PHz	100 nm	12.4 eV
NIR	300 THz	1 μm	1.24 eV
MIR	30 THz	10 μm	124 meV
FIR	3 THz	100 μm	12.4 meV
EHF	300 GHz	1 mm	1.24 meV
SHF	30 GHz	1 cm	124 μeV
UHF	3 GHz	1 dm	12.4 μeV
VHF	300 MHz	1 m	1.24 μeV
HF	30 MHz	10 m	124 neV
MF	3 MHz	100 m	12.4 neV
LF	300 kHz	1 km	1.24 neV
VLF	30 kHz	10 km	124 peV
VF/ULF	3 kHz	100 km	12.4 peV
SLF	300 Hz	1 Mm	1.24 peV
ELF	30 Hz	10 Mm	124 feV
	3 Hz	100 Mm	12.4 feV

Legend: γ = Gamma rays

HX = Hard X-rays

SX = Soft X-Rays

EUV = Extreme-ultraviolet

NUV = Near-ultraviolet

Visible light (colored bands)

NIR = Near-infrared

MIR = Mid-infrared

FIR = Far-infrared

EHF = Extremely high frequency (microwaves)

SHF = Super-high frequency (microwaves)

UHF = Ultrahigh frequency (radio waves)

VHF = Very high frequency (radio)

HF = High frequency (radio)

MF = Medium frequency (radio)

LF = Low frequency (radio)

VLF = Very low frequency (radio)

VF = Voice frequency

ULF = Ultra-low frequency (radio)

SLF = Super-low frequency (radio)

ELF = Extremely low frequency(radio)

EM radiation (the designation 'radiation' excludes static electric and magnetic and near fields) is classified by wavelength into radio, microwave, infrared, visible, ultraviolet, X-rays and gamma rays. Arbitrary electromagnetic waves can be expressed by Fourier analysis in terms of sinusoidal monochromatic waves, which in turn can each be classified into these regions of the EMR spectrum.

For certain classes of EM waves, the waveform is most usefully treated as *random*, and then spectral analysis must be done by slightly different mathematical techniques appropriate to random or stochastic processes. In such cases, the individual frequency components are represented in terms of their *power* content, and the phase information is not preserved. Such a representation is called the power spectral density of the random process. Random electromagnetic radiation requiring this kind of analysis is, for example, encountered in the interior of stars, and in certain other very wideband forms of radiation such as the Zero point wave field of the electromagnetic vacuum.

The behavior of EM radiation depends on its frequency. Lower frequencies have longer wavelengths, and higher frequencies have shorter wavelengths, and are associated with photons of higher energy. There is no fundamental limit known to these wavelengths or energies, at either end of the spectrum, although photons with energies near the Planck energy or exceeding it (far too high to have ever been observed) will require new physical theories to describe.

Soundwaves are not electromagnetic radiation. At the lower end of the electromagnetic spectrum, about 20 Hz to about 20 kHz, are frequencies that might be considered in the audio range. However, electromagnetic waves cannot be directly perceived by human ears. Sound waves are instead the oscillating compression of molecules. To be heard, electromagnetic radiation must be converted to pressure waves of the fluid in which the ear is located (whether the fluid is air, water or something else).

Interactions as a Function of Frequency

When EM radiation interacts with matter, its behavior changes qualitatively as its frequency changes.

Radio and Microwave

At radio and microwave frequencies, EMR interacts with matter largely as a bulk collection of charges which are spread out over large numbers of affected atoms. In electrical conductors, such induced bulk movement of charges (electric currents) results in absorption of the EMR, or else separations of charges that cause generation of new EMR (effective reflection of the EMR). An example is absorption or emission of radio waves by antennas, or absorption of microwaves by water or other molecules with an electric dipole moment, as for example inside a microwave oven. These interactions produce either electric currents or heat, or both.

Infrared

Like radio and microwave, infrared also is reflected by metals (as is most EMR into the ultraviolet). However, unlike lower-frequency radio and microwave radiation, Infrared EMR commonly interacts with dipoles present in single molecules, which change as atoms vibrate at the ends of a single chemical bond. It is consequently absorbed by a wide range of substances, causing them to increase in temperature as the vibrations dissipate as heat. The same process, run in reverse, causes bulk substances to radiate in the infrared spontaneously.

Visible Light

As frequency increases into the visible range, photons have enough energy to change the bond structure of some individual molecules. It is not a coincidence that this happens in the "visible range," as the mechanism of vision involves the change in bonding of a single molecule (retinal) which absorbs light in the rhodopsin in the retina of the human eye. Photosynthesis becomes possible in this range as well, for similar reasons, as a single molecule of chlorophyll is excited by a single photon. Animals that detect infrared make use of small packets of water that change temperature, in an essentially thermal process that involves many photons. For this reason, infrared, microwaves and radio waves are thought to damage molecules and biological tissue only by bulk heating, not excitation from single photons of the radiation.

Visible light is able to affect a few molecules with single photons, but usually not in a permanent or damaging way, in the absence of power high enough to increase temperature to damaging levels. However, in plant tissues that conduct photosynthesis, carotenoids act to quench electronically excited chlorophyll produced by visible light in a process called non-photochemical quenching, in order to prevent reactions that would otherwise interfere with photosynthesis at high light levels. Limited evidence indicate that some reactive oxygen species are created by visible light in skin, and that these may have some role in photoaging, in the same manner as ultraviolet A.

Ultraviolet

As frequency increases into the ultraviolet, photons now carry enough energy (about three electron volts or more) to excite certain doubly bonded molecules into permanent chemical rearrangement. In DNA, this causes lasting damage. DNA is also indirectly damaged by reactive oxygen species produced by ultraviolet A (UVA), which has energy too low to damage DNA directly. This is why ultraviolet at all wavelengths can damage DNA, and is capable of causing cancer, and (for UVB) skin burns (sunburn) that are far worse than would be produced by simple heating (temperature increase) effects. This property of causing molecular damage that is out of proportion to heating effects, is characteristic of all EMR with frequencies at the visible light range and above. These properties of high-frequency EMR are due to quantum effects that permanently damage materials and tissues at the molecular level.

At the higher end of the ultraviolet range, the energy of photons becomes large enough to impart enough energy to electrons to cause them to be liberated from the atom, in a process called photoionisation. The energy required for this is always larger than about 10 electron volts (eV) corresponding with wavelengths smaller than 124 nm (some sources suggest a more realistic cutoff of 33 eV, which is the energy required to ionize water). This high end of the ultraviolet spectrum with energies in the approximate ionization range, is sometimes called "extreme UV." Ionizing UV is strongly filtered by the Earth's atmosphere).

X-rays and Gamma Rays

Electromagnetic radiation composed of photons that carry minimum-ionization energy, or more, (which includes the entire spectrum with shorter wavelengths), is therefore termed ionizing radiation. (Many other kinds of ionizing radiation are made of non-EM particles). Electromagnetic-type ionizing radiation extends from the extreme ultraviolet to all higher frequencies and shorter wavelengths, which means that all X-rays and gamma rays qualify. These are capable of the most severe types of molecular damage, which can happen in biology to any type of biomolecule, including mutation and cancer, and often at great depths below the skin, since the higher end of the X-ray spectrum, and all of the gamma ray spectrum, penetrate matter.

Atmosphere and Magnetosphere

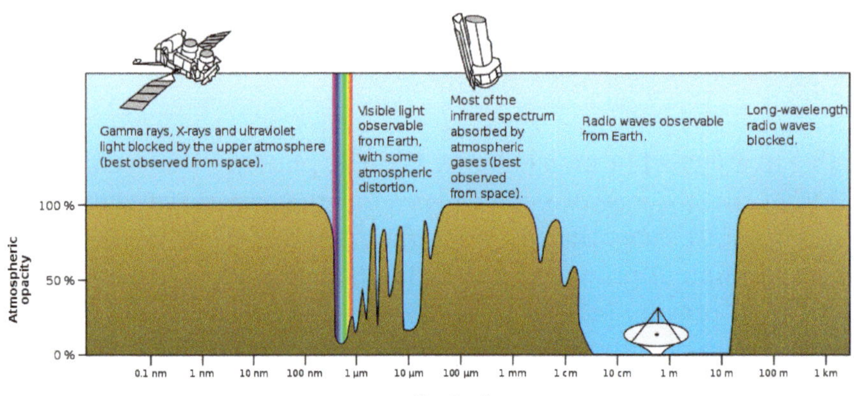

Rough plot of Earth's atmospheric absorption and scattering (or opacity) of various wavelengths of electromagnetic radiation

Most UV and X-rays are blocked by absorption first from molecular nitrogen, and then (for wavelengths in the upper UV) from the electronic excitation of dioxygen and finally ozone at the mid-range of UV. Only 30% of the Sun's ultraviolet light reaches the ground, and almost all of this is well transmitted.

Visible light is well transmitted in air, as it is not energetic enough to excite nitrogen, oxygen, or ozone, but too energetic to excite molecular vibrational frequencies of water vapor.

Absorption bands in the infrared are due to modes of vibrational excitation in water vapor. However, at energies too low to excite water vapor, the atmosphere becomes transparent again, allowing free transmission of most microwave and radio waves.

Finally, at radio wavelengths longer than 10 meters or so (about 30 MHz), the air in the lower atmosphere remains transparent to radio, but plasma in certain layers of the ionosphere begins to interact with radio waves. This property allows some longer wavelengths (100 meters or 3 MHz) to be reflected and results in shortwave radio beyond line-of-sight. However, certain ionospheric effects begin to block incoming radiowaves from space, when their frequency is less than about 10 MHz (wavelength longer than about 30 meters).

Types and Sources, Classed by Spectral Band

Radio Waves

Radio waves have the least amount of energy and the lowest frequency. When radio waves impinge upon a conductor, they couple to the conductor, travel along it and induce an electric current on the conductor surface by moving the electrons of the conducting material in correlated bunches of charge. Such effects can cover macroscopic distances in conductors (such as radio antennas), since the wavelength of radiowaves is long.

Microwaves

Microwaves are a form of electromagnetic radiation with wavelengths ranging from as long as one meter to as short as one millimeter; with frequencies between 300 MHz (0.3 GHz) and 300 GHz.

Infrared

Visible Light

Natural sources produce EM radiation across the spectrum. EM radiation with a wavelength between approximately 400 nm and 700 nm is directly detected by the human eye and perceived as visible light. Other wavelengths, especially nearby infrared (longer than 700 nm) and ultraviolet (shorter than 400 nm) are also sometimes referred to as light.

Thermal Radiation and Electromagnetic Radiation as a form of Heat

The basic structure of matter involves charged particles bound together. When electromagnetic radiation impinges on matter, it causes the charged particles to oscillate and gain energy. The ultimate fate of this energy depends on the context. It could be immediately re-radiated and appear as scattered, reflected, or transmitted radiation. It may get dissipated into other microscopic motions within the matter, coming to thermal equilibrium and manifesting itself as thermal energy in the material. With a few exceptions related to high-energy photons (such as fluorescence, harmonic generation, photochemical reactions, the photovoltaic effect for ionizing radiations at far ultraviolet, X-ray and gamma radiation), absorbed electromagnetic radiation simply deposits its energy by heating the material. This happens for infrared, microwave and radio wave radiation. Intense radio waves can thermally burn living tissue and can cook food. In addition to infrared lasers, sufficiently intense visible and ultraviolet lasers can easily set paper afire.

Ionizing radiation creates high-speed electrons in a material and breaks chemical bonds, but after these electrons collide many times with other atoms eventually most of the energy becomes thermal energy all in a tiny fraction of a second. This process makes ionizing radiation far more dangerous per unit of energy than non-ionizing radiation. This caveat also applies to UV, even though almost all of it is not ionizing, because UV can damage molecules due to electronic excitation, which is far greater per unit energy than heating effects.

Infrared radiation in the spectral distribution of a black body is usually considered a form of heat, since it has an equivalent temperature and is associated with an entropy change per unit of thermal energy. However, "heat" is a technical term in physics and thermodynamics and is often confused with thermal energy. Any type of electromagnetic energy can be transformed into thermal energy in interaction with matter. Thus, *any* electromagnetic radiation can "heat" (in the sense of increase the thermal energy termperature of) a material, when it is absorbed.

The inverse or time-reversed process of absorption is thermal radiation. Much of the thermal energy in matter consists of random motion of charged particles, and this energy can be radiated away from the matter. The resulting radiation may subsequently be absorbed by another piece of matter, with the deposited energy heating the material.

The electromagnetic radiation in an opaque cavity at thermal equilibrium is effectively a form of thermal energy, having maximum radiation entropy.

Biological Effects

Bioelectromagnetics is the study of the interactions and effects of EM radiation on living organisms. The effects of electromagnetic radiation upon living cells, including those in humans, depends upon the radiation's power and frequency. For low-frequency radiation (radio waves to visible light) the best-understood effects are those due to radiation power alone, acting through heating when radiation is absorbed. For these thermal effects, frequency is important only as it affects penetration into the organism (for example, microwaves penetrate better than infrared). Initially, it was believed that low frequency fields that were too weak to cause significant heating could not possibly have any biological effect.

Despite this opinion among researchers, evidence has accumulated that supports the existence of complex biological effects of weaker *non-thermal* electromagnetic fields, (including weak ELF magnetic fields, although the latter does not strictly qualify as EM radiation), and modulated RF and microwave fields. Fundamental mechanisms of the interaction between biological material and electromagnetic fields at non-thermal levels are not fully understood.

The World Health Organization has classified radio frequency electromagnetic radiation as Group 2B - possibly carcinogenic. This group contains possible carcinogens that have weaker evidence, at the same level as coffee and automobile exhaust. For example, epidemiological studies looking for a relationship between cell phone use and brain cancer development, have been largely inconclusive, save to demonstrate that the effect, if it exists, cannot be a large one.

At higher frequencies (visible and beyond), the effects of individual photons begin to become important, as these now have enough energy individually to directly or indirectly damage biological molecules. All UV frequences have been classed as Group 1 carcinogens by the World Health Organization. Ultraviolet radiation from sun exposure is the primary cause of skin cancer.

Thus, at UV frequencies and higher (and probably somewhat also in the visible range), electromagnetic radiation does more damage to biological systems than simple heating predicts. This is most obvious in the "far" (or "extreme") ultraviolet. UV, with X-ray and gamma radiation, are referred to as ionizing radiation due to the ability of photons of this radiation to produce ions and free radicals in materials (including living tissue). Since such radiation can severely damage life at energy levels that produce little heating, it is considered far more dangerous (in terms of damage-produced per unit of energy, or power) than the rest of the electromagnetic spectrum.

Derivation from Electromagnetic Theory

Electromagnetic waves were predicted by the classical laws of electricity and magnetism, known as Maxwell's equations. Inspection of Maxwell's equations without sources (charges or currents) results in nontrivial solutions of changing electric and magnetic fields. Beginning with Maxwell's equations in free space:

$$\nabla \cdot \mathbf{E} = 0$$

$$\nabla \times \mathbf{E} = -\frac{\partial \mathbf{B}}{\partial t}$$

$$\nabla \cdot \mathbf{B} = 0$$

$$\nabla \times \mathbf{B} = \mu_0 \epsilon_0 \frac{\partial \mathbf{E}}{\partial t}$$

where

∇ is a vector differential operator.

One solution,

$$\mathbf{E} = \mathbf{B} = \mathbf{0},$$

is trivial.

For a more useful solution, we utilize vector identities, which work for any vector, as follows:

$$\nabla \times (\nabla \times \mathbf{A}) = \nabla(\nabla \cdot \mathbf{A}) - \nabla^2 \mathbf{A}$$

The curl of earlier equation:

$$\nabla \times (\nabla \times \mathbf{E}) = \nabla \times \left(-\frac{\partial \mathbf{B}}{\partial t} \right)$$

Evaluating the left hand side:

$$\nabla \times (\nabla \times \mathbf{E}) = \nabla(\nabla \cdot \mathbf{E}) - \nabla^2 \mathbf{E} = -\nabla^2 \mathbf{E}$$

simplifying the above by using equation.

Evaluating the right hand side:

$$\nabla \times \left(-\frac{\partial \mathbf{B}}{\partial t} \right) = -\frac{\partial}{\partial t}(\nabla \times \mathbf{B}) = -\mu_0 \epsilon_0 \frac{\partial^2 \mathbf{E}}{\partial t^2}$$

The equations above are equal, so this results in a vector-valued differential equation for the electric field, namely

$$\nabla^2 \mathbf{E} = \mu_0 \epsilon_0 \frac{\partial^2 \mathbf{E}}{\partial t^2}$$

Applying a similar pattern results in similar differential equation for the magnetic field:

$$\nabla^2 \mathbf{B} = \mu_0 \epsilon_0 \frac{\partial^2 \mathbf{B}}{\partial t^2}.$$

These differential equations are equivalent to the wave equation:

$$\nabla^2 f = \frac{1}{c_0^2} \frac{\partial^2 f}{\partial t^2}$$

where

c_0 is the speed of the wave in free space and

f describes a displacement

Or more simply:

$$\Box f = 0$$

where $\Box$ is d'Alembertian:

$$\Box = \nabla^2 - \frac{1}{c_0^2}\frac{\partial^2}{\partial t^2} = \frac{\partial^2}{\partial x^2} + \frac{\partial^2}{\partial y^2} + \frac{\partial^2}{\partial z^2} - \frac{1}{c_0^2}\frac{\partial^2}{\partial t^2}$$

In the case of the electric and magnetic fields, the speed is:

$$c_0 = \frac{1}{\sqrt{\mu_0 \epsilon_0}}$$

This is the speed of light in vacuum. Maxwell's equations unified the vacuum permittivity ϵ_0, the vacuum permeability μ_0, and the speed of light itself, c_0. This relationship had been discovered by Wilhelm Eduard Weber and Rudolf Kohlrausch prior to the development of Maxwell's electrodynamics, however Maxwell was the first to produce a field theory consistent with waves traveling at the speed of light.

These are only two equations versus the original four, so more information pertains to these waves hidden within Maxwell's equations. A generic vector wave for the electric field.

$$\mathbf{E} = \mathbf{E}_0 f\left(\hat{\mathbf{k}} \cdot \mathbf{x} - c_0 t\right)$$

Here, $\mathbf{E}_0$ is the constant amplitude, f is any second differentiable function, $\hat{\mathbf{k}}$ is a unit vector in the direction of propagation, and $\mathbf{x}$ is a position vector. $f\left(\hat{\mathbf{k}} \cdot \mathbf{x} - c_0 t\right)$ is a generic solution to the wave equation. In other words,

$$\nabla^2 f\left(\hat{\mathbf{k}} \cdot \mathbf{x} - c_0 t\right) = \frac{1}{c_0^2}\frac{\partial^2}{\partial t^2} f\left(\hat{\mathbf{k}} \cdot \mathbf{x} - c_0 t\right),$$

for a generic wave traveling in the $\hat{\mathbf{k}}$ direction.

This form will satisfy the wave equation.

$$\nabla \cdot \mathbf{E} = \hat{\mathbf{k}} \cdot \mathbf{E}_0 f'\left(\hat{\mathbf{k}} \cdot \mathbf{x} - c_0 t\right) = 0$$

$$\mathbf{E} \cdot \hat{\mathbf{k}} = 0$$

The first of Maxwell's equations implies that the electric field is orthogonal to the direction the wave propagates.

$$\nabla \times \mathbf{E} = \hat{\mathbf{k}} \times \mathbf{E}_0 f'\left(\hat{\mathbf{k}} \cdot \mathbf{x} - c_0 t\right) = -\frac{\partial \mathbf{B}}{\partial t}$$

$$\mathbf{B} = \frac{1}{c_0}\hat{\mathbf{k}} \times \mathbf{E}$$

The second of Maxwell's equations yields the magnetic field. The remaining equations will be satisfied by this choice of $\mathbf{E}, \mathbf{B}$.

The electric and magnetic field waves in the far-field travel at the speed of light. They have a special restricted orientation and proportional magnitudes, $E_0 = c_0 B_0$, which can be seen immediately from the Poynting vector. The electric field, magnetic field, and direction of wave propagation are all orthogonal, and the wave propagates in the same direction as . Also, E and B far-fields in free space, which as wave solutions depend primarily on these two Maxwell equations, are in-phase with each other. This is guaranteed since the generic wave solution is first order in both space and time, and the curl operator on one side of these equations results in first-order spatial derivatives of the wave solution, while the time-derivative on the other side of the equations, which gives the other field, is first order in time, resulting in the same phase shift for both fields in each mathematical operation.

From the viewpoint of an electromagnetic wave traveling forward, the electric field might be oscillating up and down, while the magnetic field oscillates right and left. This picture can be rotated with the electric field oscillating right and left and the magnetic field oscillating down and up. This is a different solution that is traveling in the same direction. This arbitrariness in the orientation with respect to propagation direction is known as polarization. On a quantum level, it is described as photon polarization. The direction of the polarization is defined as the direction of the electric field.

More general forms of the second-order wave equations given above are available, allowing for both non-vacuum propagation media and sources. Many competing derivations exist, all with varying levels of approximation and intended applications. One very general example is a form of the electric field equation, which was factorized into a pair of explicitly directional wave equations, and then efficiently reduced into a single uni-directional wave equation by means of a simple slow-evolution approximation.

Dual Nature of Electromagnetic Radiation

- To describe the Compton observation, we need to treat radiation as localized particle, "Photon". Here, the radiation and matter interaction is particle-particle in nature.
- Wave picture is indeed required to explain interference, diffraction phenomena.
- In Compton's experiment, the particle nature is used for the description of scattering. On the other hand, the crystal is used to separate the wavelengths and this is based on diffraction (wave nature).
- Electromagnetic radiation behaves wavelike under some circumstances and particle like under other circumstances. Both pictures cannot be used simultaneously.
- This wave particle dual nature not only exists for radiation. In fact subatomic particles also follow the same.

Matter Waves

In 1924, Louis de Broglie first proposed the existence of matter waves.

He supported the wave-particle dual nature of radiation and proposed the existence of matter waves. In his hypothesis, he suggested that the wave aspect of the matter is related to its particle nature in the same quantitative way that is in case of radiation.

The total energy → related to the frequency v of the wave. $E = hv$

and the momentum p is related to the wavelength λ of the wave.

$$p = \frac{h}{\lambda}$$

It is to be noted that the particle characteristics such as E and p are related to the wave characteristics like v and λ → with Planck's constant h.

It is interesting to see that, if we calculate the de Broglie wavelength of a cricket ball of mass 0.5 Kg at a speed of 100 m/sec.

$$\lambda = \frac{h}{p} = \frac{h}{mv} = \frac{6.6\times10^{-34}\,J-\sec}{0.5Kg\times100\,{}^{m}/_{\sec}} = 1.32\times10^{-35}\,meter = 1.32\times10^{-25}\,\mathring{A}$$

On the other hand the de Broglie wavelength of an electron with kinetic energy 100 eV.

$$\lambda = \frac{6.6\times10^{-34}}{5.4\times10^{-24}\,Kg^{-m/\sec}} = 1.2\times10^{-10}\,meter = 1.2\lambda\mathring{A}$$

If one has to measure these wavelength, the instrument with aperture $a = 1\mathring{A}$.

$\frac{\lambda}{a} = 1.32\times10^{-25}$ for cricket ball

and $\frac{\lambda}{a} = 1.2$ for electron

If $\frac{\lambda}{a} \to 0$, the diffraction effects will be negligible and measurement is not possible.

So, what is the limit of measurement?

Uncertainty Principle

As discussed earlier, that in classical mechanics, the equation of the motion with given forces can be solved by knowing the position and the momentum of the particle for all values of t.

To measure these observables one need to disturb the system. However for macroscopic particle, this disturbance is small enough to ignore.

But for microscopic particle, simultaneous measurement can not be done beyond the accuracy governed by Heisenberg uncertainty principle.

Precision of the measurement is limited by

$$\Delta p_x \Delta x \geq \hbar/2 \qquad [\hbar = h/2\pi]$$

where Δp_x is the uncertainty of knowing the component of momentum p in x-direction and Δx is the uncertainty in locating the particle at coordinate x.

The second part is,

$$\Delta E \Delta t \geq \hbar/2$$

where ΔE and Δt are the uncertainties of Energy E and time t.

This concept helps to understand the limitations of the classical theory and provides the regime where the quantum mechanics is valid.

Failures of Bohr Model

Bohr model was a major step toward understanding the quantum theory of the atom - not in fact a correct description of the nature of electron orbits.

Some of the shortcomings of the model are:

- Fails describe why certain spectral lines are brighter than others => no mechanism for calculating transition probabilities.
- Violates the uncertainty principal which dictates that position and momentum cannot be simultaneously determined.

Bohr model gives a basic conceptual model of electrons orbits and energies. The precise details can only be solved using the Schrödinger equation.

From the Bohr model, the linear momentum of the electron is

$$p = mv = m\left(\frac{Ze^2}{4\pi\varepsilon_0 n\hbar}\right) = \frac{n\hbar}{r}$$

However, we know from Hiesenberg Uncertainty Principle, that

$$\Delta p \sim \frac{\hbar}{\Delta x} \sim \frac{\hbar}{r}$$

Comparing Equation given above, we get => $p \sim nDp$

This shows that the magnitude of p is undefined except when n is large.

- *Bohr model only valid when we approach the classical limit at large n.*
- *Must therefore use full quantum mechanical treatment to model electron in H* atom.
- One of Bohr's postulates was the angular momentum of the electron is quantized, but there was no explanation why the restriction occurred
- de Broglie assumed that the electron orbit would be stable only if it contained an integral number of electron wavelengths

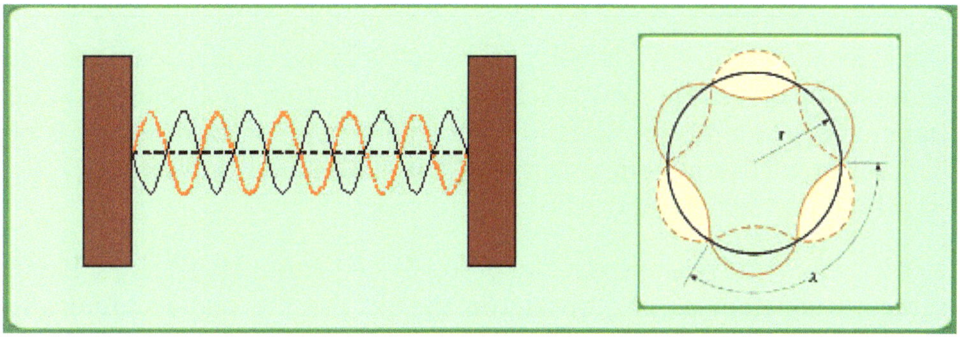

Idea is very similar to vibrations of a string tied both sides with wall, as shown in the figure above.

Two nodes at the two ends.

If these nodes join together, it becomes a circle and standing waves will generate nodes and antinodes.

- In this figure, three complete wavelengths are contained in the circumference of the orbit
- In general, the circumference must equal some integer number of wavelengths - $2 p r = n \lambda$ $n = 1, 2, \ldots$
- The expression for the de Broglie wavelength can be included in the circumference calculation- $me\ v\ r = n \hbar$

This is the same quantization of angular momentum that Bohr imposed in his original theory

- This was the first convincing argument that the wave nature of matter was at the heart of the behavior of atomic systems
- By applying wave theory to the electrons in an atom, de Broglie was able to explain the appearance of integers in Bohr's equations as a natural consequence of standing wave patterns
- Schrödinger's wave equation was subsequently applied to atomic systems

Schrödinger Equation

In quantum mechanics, the Schrödinger equation is a mathematical equation that describes the evolution over time of a physical system in which quantum effects, such as wave–particle duality, are significant. The equation is a mathematical formulation for studying quantum mechanical systems. It is considered a central result in the study of quantum system and its derivation was a significant landmark in developing the theory of quantum mechanics. It was named after Erwin Schrödinger, who derived the equation in 1925 and published it in 1926, forming the basis for his work that resulted in Schrödinger being awarded the Nobel Prize in Physics in 1933. The equation is a type of differential equation known as a wave-equation, which serves as a mathematical model of the movement of waves.

In classical mechanics, Newton's second law (**F** = *m***a**) is used to make a mathematical prediction as to what path a given system will take following a set of known initial conditions. In quantum mechanics, the analogue of Newton's law is Schrödinger's equation for a quantum system (usually atoms, molecules, and subatomic particles whether free, bound, or localised). It is not a simple algebraic equation, but in general a linear partial differential equation, describing the time-evolution of the system's wave function (also called a "state function").

The concept of a wavefunction is a fundamental postulate of quantum mechanics. Using these postulates, Schrödinger's equation can be derived from the fact that the time-evolution operator must be unitary and must therefore be generated by the exponential of a self-adjoint operator, which is the quantum Hamiltonian. This derivation is explained below.

In the Copenhagen interpretation of quantum mechanics, the wave function is the most complete description that can be given of a physical system. Solutions to Schrödinger's equation describe not only molecular, atomic, and subatomic systems, but also macroscopic systems, possibly even the whole universe. Schrödinger's equation is central to all applications of quantum mechanics including quantum field theory which combines special relativity with quantum mechanics. Theories of quantum gravity, such as string theory, also do not modify Schrödinger's equation.

The Schrödinger equation is not the only way to study quantum mechanical systems and make predictions, as there are other quantum mechanical formulations such as matrix mechanics, introduced by Werner Heisenberg, and path integral formulation, developed chiefly by Richard Feynman. Paul Dirac incorporated matrix mechanics and the Schrödinger equation into a single formulation.

Equation

Time-dependent Equation

The form of the Schrödinger equation depends on the physical situation. The most general form is the time-dependent Schrödinger equation, which gives a description of a system evolving with time:

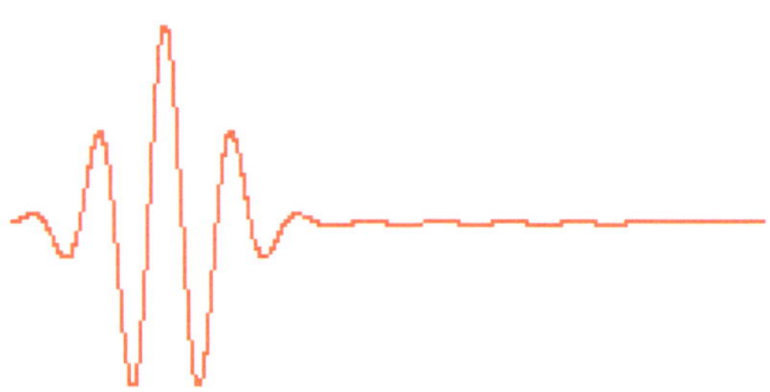

A wave function that satisfies the nonrelativistic Schrödinger equation with $V = 0$. In other words, this corresponds to a particle traveling freely through empty space. The real part of the wave function is plotted here.

Time-dependent Schrödinger equation (general)

$$i\hbar\frac{\partial}{\partial t}\Psi(\mathbf{r},t) = \hat{H}\Psi(\mathbf{r},t)$$

where i is the imaginary unit, $\hbar$ is the reduced Planck constant which is: $\hbar = \frac{h}{2\pi}$ the symbol $\frac{\partial}{\partial t}$ indicates a partial derivative with respect to time t, Ψ (the Greek letter psi) is the wave function of the quantum system, r and t are the position vector and time respectively, and $\hat{H}$ is the Hamiltonian operator (which characterizes the total energy of any given wave function and takes different forms depending on the situation).

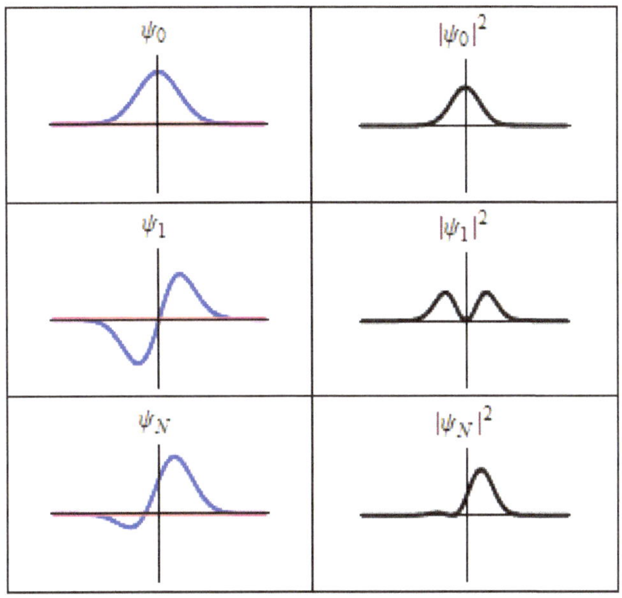

Each of these three rows is a wave function which satisfies the time-dependent Schrödinger equation for a harmonic oscillator. Left: The real part (blue) and imaginary part (red) of the wave function. Right: The probability distribution of finding the particle with this wave function at a given position. The top two rows are examples of stationary states, which correspond to standing waves. The bottom row is an example of a state which is *not* a stationary state. The right column illustrates why stationary states are called "stationary".

The most famous example is the nonrelativistic Schrödinger equation for a single particle moving in an electric field (but not a magnetic field):

Time-dependent Schrödinger equation *(single nonrelativistic particle)*

$$i\hbar \frac{\partial}{\partial t}\Psi(\mathbf{r},t) = \left[\frac{-\hbar^2}{2\mu}\nabla^2 + V(\mathbf{r},t)\right]\Psi(\mathbf{r},t)$$

where μ is the particle's "reduced mass", V is its potential energy, ∇^2 is the Laplacian (a differential operator), and Ψ is the wave function (more precisely, in this context, it is called the "position-space wave function"). In plain language, it means "total energy equals kinetic energy plus potential energy", but the terms take unfamiliar forms for reasons explained below.

Given the particular differential operators involved, this is a linear partial differential equation. It is also a diffusion equation, but unlike the heat equation, this one is also a wave equation given the imaginary unit present in the transient term.

The term *"Schrödinger equation"* can refer to both the general equation (first box above), or the specific nonrelativistic version (second box above and variations thereof). The general equation is indeed quite general, used throughout quantum mechanics, for everything from the Dirac equation to quantum field theory, by plugging in various complicated expressions for the Hamiltonian. The specific nonrelativistic version is a simplified approximation to reality, which is quite accurate in many situations, but very inaccurate in others.

To apply the Schrödinger equation, the Hamiltonian operator is set up for the system, accounting for the kinetic and potential energy of the particles constituting the system, then inserted into the Schrödinger equation. The resulting partial differential equation is solved for the wave function, which contains information about the system.

Time-independent Equation

The time-dependent Schrödinger equation described above predicts that wave functions can form standing waves, called stationary states (also called "orbitals", as in atomic orbitals or molecular orbitals). These states are important in their own right, and if the stationary states are classified and understood, then it becomes easier to solve the time-dependent Schrödinger equation for *any* state. Stationary states can also be described by a simpler form of the Schrödinger equation, the *time-independent Schrödinger equation.* (This is only used when the Hamiltonian itself is not dependent on time explicitly. However, even in this case the total wave function still has a time dependency.)

Time-independent Schrödinger equation (*general*)

$$\hat{\mathrm{H}}\Psi = E\Psi$$

In words, the equation states:

> *When the Hamiltonian operator acts on a certain wave function Ψ, and the result is proportional to the same wave function Ψ, then Ψ is a stationary state, and the proportionality constant, E, is the energy of the state Ψ.*

The time-independent Schrödinger equation is discussed further below. In linear algebra terminology, this equation is an eigenvalue equation.

As before, the most famous manifestation is the nonrelativistic Schrödinger equation for a single particle moving in an electric field (but not a magnetic field):

Time-independent Schrödinger equation (*single nonrelativistic particle*)

$$\left[\frac{-\hbar^2}{2\mu}\nabla^2 + V(\mathbf{r})\right]\Psi(\mathbf{r}) = E\Psi(\mathbf{r})$$

with definitions as above.

Derivation

In the modern understanding of quantum mechanics, Schrödinger's equation may be derived as follows. If the wave-function at time t is given by $\Psi(t)$, then by the linearity of quantum mechanics the wave-function at time t' must be given by $\Psi(t') = U(t',t)\Psi(t)$, where $U(t',t)$ is a linear operator. Since time-evolution must preserve the norm of the wave-function, it follows that $U(t',t)$ must be a member of the unitary group of operators acting on wave-functions. We also know that when $t' = t$, we must have $U(t,t) = 1$. Therefore, expanding the operator $U(t',t)$ for t' close to t, we can write $U(t',t) = 1 - iH(t' - t)$ where H is a Hermitian operator. This follows from the fact that the Lie algebra corresponding to the unitary group comprises Hermitian operators. Taking the limit as the time-difference $t' - t$ becomes very small, we obtain Schrödinger's equation.

So far, H is only an abstract Hermitian operator. However using the correspondence principle it is possible to show that, in the classical limit, the expectation value of H is indeed the classical energy. The correspondence principle does not completely fix the form of the quantum Hamiltonian due to the uncertainty principle and therefore the precise form of the quantum Hamiltonian must be fixed empirically.

Implications

The Schrödinger equation and its solutions introduced a breakthrough in thinking about physics. Schrödinger's equation was the first of its type, and solutions led to consequences that were very unusual and unexpected for the time.

Total, Kinetic, and Potential Energy

The *overall* form of the equation is *not* unusual or unexpected, as it uses the principle of the conservation of energy. The terms of the nonrelativistic Schrödinger equation can be interpreted as total energy of the system, equal to the system kinetic energy plus the system potential energy. In this respect, it is just the same as in classical physics.

Quantization

The Schrödinger equation predicts that if certain properties of a system are measured, the result may be *quantized*, meaning that only specific discrete values can occur. One example is *energy quantization*: the energy of an electron in an atom is always one of the quantized energy levels, a fact discovered via atomic spectroscopy. Another example is quantization of angular momentum. This was an *assumption* in the earlier Bohr model of the atom, but it is a *prediction* of the Schrödinger equation.

Another result of the Schrödinger equation is that not every measurement gives a quantized result in quantum mechanics. For example, position, momentum, time, and (in some situations) energy can have any value across a continuous range.

Measurement and Uncertainty

In classical mechanics, a particle has, at every moment, an exact position and an exact momentum. These values change deterministically as the particle moves according to Newton's laws. Under the Copenhagen interpretation of quantum mechanics, particles do not have exactly determined properties, and when they are measured, the result is randomly drawn from a probability distribution. The Schrödinger equation predicts what the probability distributions are, but fundamentally cannot predict the exact result of each measurement.

The Heisenberg uncertainty principle is the statement of the inherent measurement uncertainty in quantum mechanics. It states that the more precisely a particle's position is known, the less precisely its momentum is known, and vice versa.

The Schrödinger equation describes the (deterministic) evolution of the wave function of a particle. However, even if the wave function is known exactly, the result of a specific measurement on the wave function is uncertain.

Quantum Tunneling

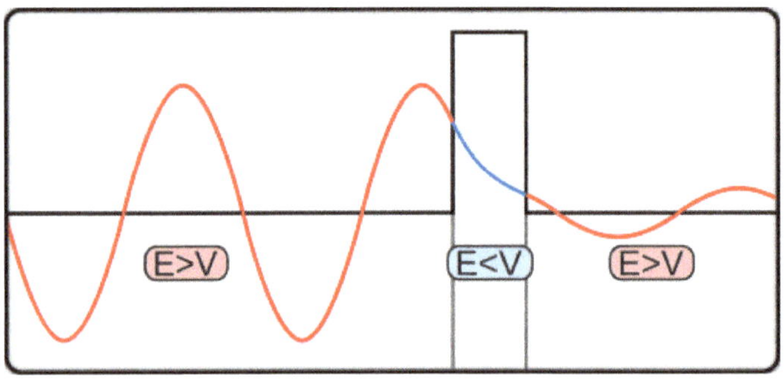

Quantum tunneling through a barrier. A particle coming from the left does not have enough energy to climb the barrier. However, it can sometimes "tunnel" to the other side

In classical physics, when a ball is rolled slowly up a large hill, it will come to a stop and roll back, because it doesn't have enough energy to get over the top of the hill to the other side. However, the Schrödinger equation predicts that there is a small probability that the ball will get to the other side of the hill, even if it has too little energy to reach the top. This is called quantum tunneling. It is related to the distribution of energy: although the ball's assumed position seems to be on one side of the hill, there is a chance of finding it on the other side.

Particles as Waves

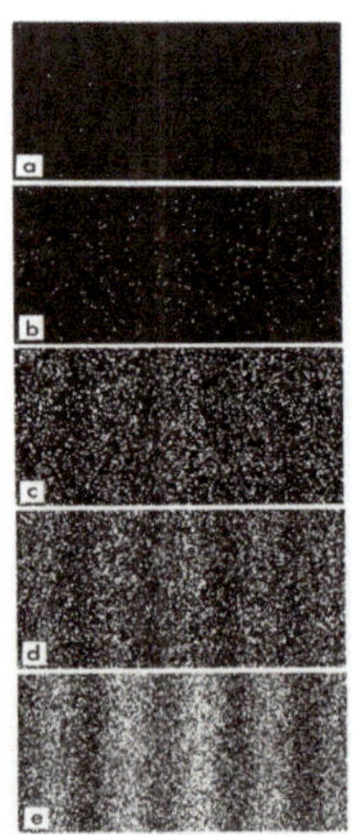

A double slit experiment showing the accumulation of electrons on a screen as time passes

The nonrelativistic Schrödinger equation is a type of partial differential equation called a wave equation. Therefore, it is often said particles can exhibit behavior usually attributed to waves. In some modern interpretations this description is reversed – the quantum state, i.e. wave, is the only genuine physical reality, and under the appropriate conditions it can show features of particle-like behavior. However, Ballentine shows that such an interpretation has problems. Ballentine points out that whilst it is arguable to associate a physical wave with a single particle, there is still only *one* Schrödinger wave equation for many particles. He points out:

> *"If a physical wave field were associated with a particle, or if a particle were identified with a wave packet, then corresponding to N interacting particles there should be N interacting waves in ordinary three-dimensional space. But according to (4.6) that is not the case; instead there is one "wave" function in an abstract 3N-dimensional configuration space. The misinterpretation of psi as a physical wave in ordinary space is possible only because the most common applications of quantum mechanics are to one-particle states, for which configuration space and ordinary space are isomorphic."*

Two-slit diffraction is a famous example of the strange behaviors that waves regularly display, that are not intuitively associated with particles. The overlapping waves from the two slits cancel each other out in some locations, and reinforce each other in other locations, causing a complex pattern to emerge. Intuitively, one would not expect this pattern from firing a single particle at the slits, because the particle should pass through one slit or the other, not a complex overlap of both.

However, since the Schrödinger equation is a wave equation, a single particle fired through a double-slit *does* show this same pattern. The experiment must be repeated many times for the complex pattern to emerge. Although this is counterintuitive, the prediction is correct; in particular, electron diffraction and neutron diffraction are well understood and widely used in science and engineering.

Related to diffraction, particles also display superposition and interference.

The superposition property allows the particle to be in a quantum superposition of two or more quantum states at the same time. However, it is noted that a "quantum state" in quantum mechanics means the *probability* that a system will be, for example at a position x, not that the system will actually be at position x. It does not imply that the particle itself may be in two classical states at once. Indeed, QM is generally unable to assign values for properties prior to measurement at all.

Multiverse

In Dublin in 1952 Erwin Schrödinger gave a lecture in which at one point he jocularly warned his audience that what he was about to say might "seem lunatic". It was that, when his Nobel equations seem to be describing several different histories, they are "not alternatives but all really happen simultaneously". This is the earliest known reference to the multiverse.

Interpretation of the Wave Function

The Schrödinger equation provides a way to calculate the wave function of a system and how it changes dynamically in time. However, the Schrödinger equation does not directly say *what*, exactly, the wave function is. Interpretations of quantum mechanics address questions such as what the relation is between the wave function, the underlying reality, and the results of experimental measurements.

An important aspect is the relationship between the Schrödinger equation and wavefunction collapse. In the oldest Copenhagen interpretation, particles follow the Schrödinger equation *except* during wavefunction collapse, during which they behave entirely differently. The advent of quantum decoherence theory allowed alternative approaches (such as the Everett many-worlds interpretation and consistent histories), wherein the Schrödinger equation is *always* satisfied, and wavefunction collapse should be explained as a consequence of the Schrödinger equation.

Historical Background and Development

Erwin Schrödinger

Following Max Planck's quantization of light, Albert Einstein interpreted Planck's quanta to be photons, particles of light, and proposed that the energy of a photon is proportional to its frequency, one of the first signs of wave–particle duality. Since energy and momentum are related in the same way as frequency and wavenumber in special relativity, it followed that the momentum p of a photon is inversely proportional to its wavelength λ, or proportional to its wavenumber k.

$$p = \frac{h}{\lambda} = \hbar k$$

where h is Planck's constant. Louis de Broglie hypothesized that this is true for all particles, even particles which have mass such as electrons. He showed that, assuming that the matter waves propagate along with their particle counterparts, electrons form standing waves, meaning that only certain discrete rotational frequencies about the nucleus of an atom are allowed. These quantized orbits correspond to discrete energy levels, and de Broglie reproduced the Bohr model formula for the energy levels. The Bohr model was based on the assumed quantization of angular momentum L according to:

$$L = n\frac{h}{2\pi} = n\hbar.$$

According to de Broglie the electron is described by a wave and a whole number of wavelengths must fit along the circumference of the electron's orbit:

$$n\lambda = 2\pi r.$$

This approach essentially confined the electron wave in one dimension, along a circular orbit of radius r.

In 1921, prior to de Broglie, Arthur C. Lunn at the University of Chicago had used the same argument based on the completion of the relativistic energy–momentum 4-vector to derive what we now call the de Broglie relation. Unlike de Broglie, Lunn went on to formulate the differential equation now known as the Schrödinger equation, and solve for its energy eigenvalues for the hydrogen atom. Unfortunately the paper was rejected by the Physical Review, as recounted by Kamen.

Following up on de Broglie's ideas, physicist Peter Debye made an offhand comment that if particles behaved as waves, they should satisfy some sort of wave equation. Inspired by Debye's remark, Schrödinger decided to find a proper 3-dimensional wave equation for the electron. He was guided by William R. Hamilton's analogy between mechanics and optics, encoded in the observation that the zero-wavelength limit of optics resembles a mechanical system—the trajectories of light rays become sharp tracks that obey Fermat's principle, an analog of the principle of least action. A modern version of his reasoning is reproduced below. The equation he found is:

$$i\hbar\frac{\partial}{\partial t}\Psi(\mathbf{r},t) = -\frac{\hbar^2}{2m}\nabla^2\Psi(\mathbf{r},t) + V(\mathbf{r})\Psi(\mathbf{r},t).$$

However, by that time, Arnold Sommerfeld had refined the Bohr model with relativistic corrections. Schrödinger used the relativistic energy momentum relation to find what is now known as the Klein–Gordon equation in a Coulomb potential (in natural units):

$$\left(E + \frac{e^2}{r}\right)^2 \psi(x) = -\nabla^2\psi(x) + m^2\psi(x).$$

He found the standing waves of this relativistic equation, but the relativistic corrections disagreed with Sommerfeld's formula. Discouraged, he put away his calculations and secluded himself in an isolated mountain cabin in December 1925.

While at the cabin, Schrödinger decided that his earlier nonrelativistic calculations were novel enough to publish, and decided to leave off the problem of relativistic corrections for the future. Despite the difficulties in solving the differential equation for hydrogen (he had sought help from his friend the mathematician Hermann Weyl) Schrödinger showed that his nonrelativistic version of the wave equation produced the correct spectral energies of hydrogen in a paper published in 1926. In the equation, Schrödinger computed the hydrogen spectral series by treating a hydrogen atom's electron as a wave $\Psi(x, t)$, moving in a potential well V, created by the proton. This computation accurately reproduced the energy levels of the Bohr model. In a paper, Schrödinger himself explained this equation as follows:

> "The already ... mentioned psi-function.... is now the means for predicting probability of measurement results. In it is embodied the momentarily attained sum of theoretically based future expectation, somewhat as laid down in a catalog."
>
> — Erwin Schrödinger

This 1926 paper was enthusiastically endorsed by Einstein, who saw the matter-waves as an intuitive depiction of nature, as opposed to Heisenberg's matrix mechanics, which he considered overly formal.

The Schrödinger equation details the behavior of Ψ but says nothing of its *nature*. Schrödinger tried to interpret it as a charge density in his fourth paper, but he was unsuccessful. In 1926, just a few days after Schrödinger's fourth and final paper was published, Max Born successfully interpreted Ψ as the probability amplitude, whose absolute square is equal to probability density. Schrödinger, though, always opposed a statistical or probabilistic approach, with its associated discontinuities—much like Einstein, who believed that quantum mechanics was a statistical approximation to an underlying deterministic theory—and never reconciled with the Copenhagen interpretation.

Louis de Broglie in his later years proposed a real valued wave function connected to the complex wave function by a proportionality constant and developed the De Broglie–Bohm theory.

The Wave Equation for Particles

The Schrödinger equation is a diffusion equation, the *solutions* are functions which describe wave-like motions. Wave equations in physics can normally be derived from other physical laws – the wave equation for mechanical vibrations on strings and in matter can be derived from Newton's laws – where the wave function represents the displacement of matter, and electromagnetic waves from Maxwell's equations, where the wave functions are electric and magnetic fields. The basis for Schrödinger's equation, on the other hand, is the energy of the system and a separate postulate of quantum mechanics: the wave function is a description of the system. The Schrödinger equation is therefore a new concept in itself; as Feynman put it:

> "Where did we get that (equation) from? Nowhere. It is not possible to derive it from anything you know. It came out of the mind of Schrödinger."
>
> — Richard Feynman

The foundation of the equation is structured to be a linear differential equation based on classical energy conservation, and consistent with the De Broglie relations. The solution is the wave function ψ, which contains all the information that can be known about the system. In the Copenhagen interpretation, the modulus of ψ is related to the probability the particles are in some spatial configuration at some instant of time. Solving the equation for ψ can be used to predict how the particles will behave under the influence of the specified potential and with each other.

The Schrödinger equation was developed principally from the De Broglie hypothesis, a wave equation that would describe particles, and can be constructed.

Consistency with Energy Conservation

The total energy E of a particle is the sum of kinetic energy T and potential energy V, this sum is also the frequent expression for the Hamiltonian H in classical mechanics:

$$E = T + V = H$$

Explicitly, for a particle in one dimension with position x, mass m and momentum p, and potential energy V which generally varies with position and time t:

$$E = \frac{p^2}{2m} + V(x,t) = H.$$

For three dimensions, the position vector **r** and momentum vector **p** must be used:

$$E = \frac{\mathbf{p}\cdot\mathbf{p}}{2m} + V(\mathbf{r},t) = H$$

This formalism can be extended to any fixed number of particles: the total energy of the system is then the total kinetic energies of the particles, plus the total potential energy, again the Hamiltonian. However, there can be interactions between the particles (an N-body problem), so the potential energy V can change as the spatial configuration of particles changes, and possibly with time. The potential energy, in general, is *not* the sum of the separate potential energies for each particle, it is a function of all the spatial positions of the particles. Explicitly:

$$E = \sum_{n=1}^{N} \frac{\mathbf{p}_n\cdot\mathbf{p}_n}{2m_n} + V(\mathbf{r}_1, \mathbf{r}_2 \cdots \mathbf{r}_N, t) = H$$

Linearity

The simplest wavefunction is a plane wave of the form:

$$\Psi(\mathbf{r},t) = Ae^{i(\mathbf{k}\cdot\mathbf{r}-\omega t)}$$

where the A is the amplitude, k the wavevector, and ω the angular frequency, of the plane wave. In general, physical situations are not purely described by plane waves, so for generality the superposition principle is required; any wave can be made by superposition of sinusoidal plane waves. So if the equation is linear, a linear combination of plane waves is also an allowed solution. Hence a necessary and separate requirement is that the Schrödinger equation is a linear differential equation.

For discrete k the sum is a superposition of plane waves:

$$\Psi(\mathbf{r},t) = \sum_{n=1}^{\infty} A_n e^{i(\mathbf{k}_n\cdot\mathbf{r}-\omega_n t)}$$

for some real amplitude coefficients A_n, and for continuous k the sum becomes an integral, the Fourier transform of a momentum space wavefunction:

$$\Psi(\mathbf{r},t) = \frac{1}{(\sqrt{2\pi})^3}\int \Phi(\mathbf{k})e^{i(\mathbf{k}\cdot\mathbf{r}-\omega t)}d^3\mathbf{k}$$

where $d^3k = dk_x dk_y dk_z$ is the differential volume element in k-space, and the integrals are taken over all k-space. The momentum wavefunction Φ(k) arises in the integrand since the position and momentum space wavefunctions are Fourier transforms of each other.

Consistency with the De Broglie Relations

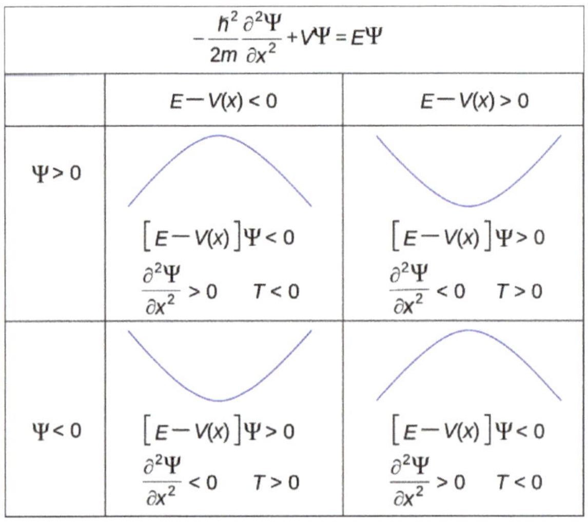

Diagrammatic summary of the quantities related to the wavefunction, as used in De broglie's hypothesis and development of the Schrödinger equation

Einstein's light quanta hypothesis (1905) states that the energy E of a photon is proportional to the frequency v (or angular frequency, $\omega = 2\pi v$) of the corresponding quantum wavepacket of light:

$$E = h\nu = \hbar\omega$$

Likewise De Broglie's hypothesis (1924) states that any particle can be associated with a wave, and that the momentum p of the particle is inversely proportional to the wavelength λ of such a wave (or proportional to the wavenumber, $k = \frac{2\pi}{\lambda}$), in one dimension, by:

$$p = \frac{h}{\lambda} = \hbar k\,,$$

while in three dimensions, wavelength λ is related to the magnitude of the wavevector k:

$$\mathbf{p} = \hbar\mathbf{k}, \quad |\mathbf{k}| = \frac{2\pi}{\lambda}.$$

The Planck–Einstein and de Broglie relations illuminate the deep connections between energy with time, and space with momentum, and express wave–particle duality. In practice, natural units comprising $\hbar = 1$ are used, as the De Broglie *equations* reduce to *identities*: allowing momentum, wavenumber, energy and frequency to be used interchangeably, to prevent duplication of quantities, and reduce the number of dimensions of related quantities. For familiarity SI units are still used in this article.

Schrödinger's insight, late in 1925, was to express the phase of a plane wave as a complex phase factor using these relations:

$$\Psi = Ae^{i(\mathbf{k}\cdot\mathbf{r}-\omega t)} = Ae^{i(\mathbf{p}\cdot\mathbf{r}-Et)/\hbar}$$

and to realize that the first order partial derivatives were:

with respect to space:

$$\nabla\Psi = \frac{i}{\hbar}\mathbf{p}Ae^{i(\mathbf{p}\cdot\mathbf{r}-Et)/\hbar} = \frac{i}{\hbar}\mathbf{p}\Psi$$

with respect to time:

$$\frac{\partial\Psi}{\partial t} = -\frac{iE}{\hbar}Ae^{i(\mathbf{p}\cdot\mathbf{r}-Et)/\hbar} = -\frac{iE}{\hbar}\Psi$$

Another postulate of quantum mechanics is that all observables are represented by linear Hermitian operators which act on the wavefunction, and the eigenvalues of the operator are the values the observable takes. The previous derivatives are consistent with the energy operator, corresponding to the time derivative,

$$\hat{E}\Psi = i\hbar\frac{\partial}{\partial t}\Psi = E\Psi$$

where E are the energy eigenvalues, and the momentum operator, corresponding to the spatial derivatives (the gradient ∇),

$$\hat{\mathbf{p}}\Psi = -i\hbar\nabla\Psi = \mathbf{p}\Psi$$

where p is a vector of the momentum eigenvalues. In the above, the "hats" (^) indicate these observables are operators, not simply ordinary numbers or vectors. The energy and momentum operators are *differential operators*, while the potential energy function V is just a multiplicative factor.

Substituting the energy and momentum operators into the classical energy conservation equation obtains the operator:

$$E = \frac{\mathbf{p}\cdot\mathbf{p}}{2m} + V \quad\rightarrow\quad \hat{E} = \frac{\hat{\mathbf{p}}\cdot\hat{\mathbf{p}}}{2m} + V$$

so in terms of derivatives with respect to time and space, acting this operator on the wavefunction Ψ immediately led Schrödinger to his equation:

$$i\hbar\frac{\partial\Psi}{\partial t} = -\frac{\hbar^2}{2m}\nabla^2\Psi + V\Psi$$

Wave–particle duality can be assessed from these equations as follows. The kinetic energy T is related to the square of momentum p. As the particle's momentum increases, the kinetic energy increases more rapidly, but since the wavenumber |k| increases the wavelength λ decreases. In terms of ordinary scalar and vector quantities (not operators):

$$\mathbf{p}\cdot\mathbf{p} \propto \mathbf{k}\cdot\mathbf{k} \propto T \propto \frac{1}{\lambda^2}$$

The kinetic energy is also proportional to the second spatial derivatives, so it is also proportional to the magnitude of the *curvature* of the wave, in terms of operators:

$$\hat{T}\Psi = \frac{-\hbar^2}{2m}\nabla\cdot\nabla\Psi \propto \nabla^2\Psi.$$

As the curvature increases, the amplitude of the wave alternates between positive and negative more rapidly, and also shortens the wavelength. So the inverse relation between momentum and wavelength is consistent with the energy the particle has, and so the energy of the particle has a connection to a wave, all in the same mathematical formulation.

Wave and Particle Motion

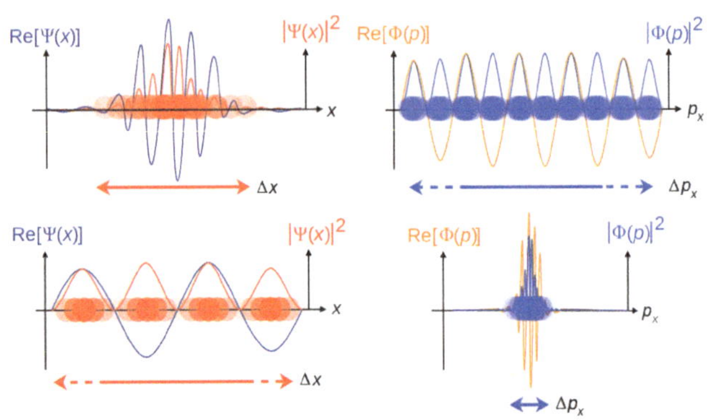

Increasing levels of wavepacket localization, meaning the particle has a more localized position.

Perfect localization

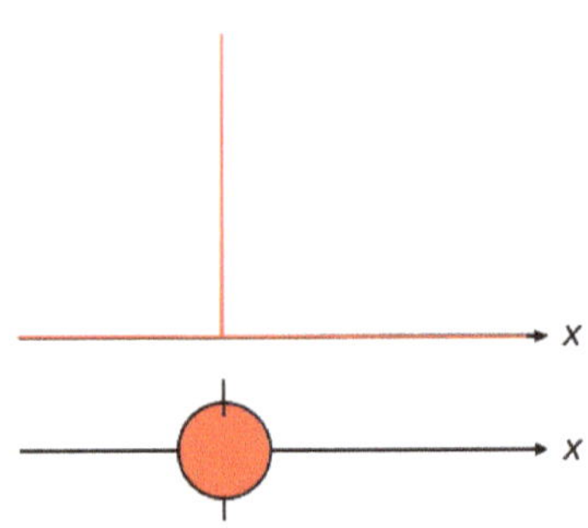

In the limit ħ → 0, the particle's position and momentum become known exactly. This is equivalent to the classical particle

Schrödinger required that a wave packet solution near position **r** with wavevector near k will move along the trajectory determined by classical mechanics for times short enough for the spread in k (and hence in velocity) not to substantially increase the spread in **r**. Since, for a given spread in k, the spread in velocity is proportional to Planck's constant ħ, it is sometimes said that in the limit as ħ approaches zero, the equations of classical mechanics are restored from quantum mechanics. Great care is required in how that limit is taken, and in what cases.

The limiting short-wavelength is equivalent to $\hbar$ tending to zero because this is limiting case of increasing the wave packet localization to the definite position of the particle. Using the Heisenberg uncertainty principle for position and momentum, the products of uncertainty in position and momentum become zero as $\hbar \rightarrow 0$:

$$\sigma(x)\sigma(p_x) \geqslant \frac{\hbar}{2} \quad \rightarrow \quad \sigma(x)\sigma(p_x) \geqslant 0$$

where σ denotes the (root mean square) measurement uncertainty in x and p_x (and similarly for the y and z directions) which implies the position and momentum can only be known to arbitrary precision in this limit.

The Schrödinger equation in its general form

$$i\hbar\frac{\partial}{\partial t}\Psi(\mathbf{r},t) = \hat{H}\Psi(\mathbf{r},t)$$

is closely related to the Hamilton–Jacobi equation (HJE)

$$\frac{\partial}{\partial t}S(q_i,t) = H\left(q_i, \frac{\partial S}{\partial q_i}, t\right)$$

where S is action and H is the Hamiltonian function (not operator). Here the generalized coordinates q_i for $i = 1, 2, 3$ (used in the context of the HJE) can be set to the position in Cartesian coordinates as $\mathbf{r} = (q_1, q_2, q_3) = (x, y, z)$.

Substituting

$$\Psi = \sqrt{\rho(\mathbf{r},t)}e^{iS(\mathbf{r},t)/\hbar}$$

where ρ is the probability density, into the Schrödinger equation and then taking the limit $\hbar \rightarrow 0$ in the resulting equation, yields the Hamilton–Jacobi equation.

The implications are:

- The motion of a particle, described by a (short-wavelength) wave packet solution to the Schrödinger equation, is also described by the Hamilton–Jacobi equation of motion.
- The Schrödinger equation includes the wavefunction, so its wave packet solution implies the position of a (quantum) particle is fuzzily spread out in wave fronts. On the contrary, the Hamilton–Jacobi equation applies to a (classical) particle of definite position and momentum, instead the position and momentum at all times (the trajectory) are deterministic and can be simultaneously known.

Nonrelativistic Quantum Mechanics

The quantum mechanics of particles without accounting for the effects of special relativity, for example particles propagating at speeds much less than light, is known as nonrelativistic quantum mechanics. Following are several forms of Schrödinger's equation in this context for different situations: time independence and dependence, one and three spatial dimensions, and one and N particles.

In actuality, the particles constituting the system do not have the numerical labels used in theory. The language of mathematics forces us to label the positions of particles one way or another, otherwise there would be confusion between symbols representing which variables are for which particle.

Time Independent

If the Hamiltonian is not an explicit function of time, the equation is separable into a product of spatial and temporal parts. In general, the wavefunction takes the form:

$$\Psi(\text{space coords}, t) = \psi(\text{space coords})\tau(t).$$

where ψ(space coords) is a function of all the spatial coordinate(s) of the particle(s) constituting the system only, and $\tau(t)$ is a function of time only.

Substituting for ψ into the Schrödinger equation for the relevant number of particles in the relevant number of dimensions, solving by separation of variables implies the general solution of the time-dependent equation has the form:

$$\Psi(\text{space coords}, t) = \psi(\text{space coords})e^{-iEt/\hbar}.$$

Since the time dependent phase factor is always the same, only the spatial part needs to be solved for in time independent problems. Additionally, the energy operator $\hat{E} = i\hbar\frac{\partial}{\partial t}$ can always be replaced by the energy eigenvalue E, thus the time independent Schrödinger equation is an eigenvalue equation for the Hamiltonian operator:

$$\hat{H}\psi = E\psi$$

This is true for any number of particles in any number of dimensions (in a time independent potential). This case describes the standing wave solutions of the time-dependent equation, which are the states with definite energy (instead of a probability distribution of different energies). In physics, these standing waves are called "stationary states" or "energy eigenstates"; in chemistry they are called "atomic orbitals" or "molecular orbitals". Superpositions of energy eigenstates change their properties according to the relative phases between the energy levels.

The energy eigenvalues from this equation form a discrete spectrum of values, so mathematically energy must be quantized. More specifically, the energy eigenstates form a basis – any wavefunction may be written as a sum over the discrete energy states or an integral over continuous energy states, or more generally as an integral over a measure. This is the spectral theorem in mathematics, and in a finite state space it is just a statement of the completeness of the eigenvectors of a Hermitian matrix.

One-dimensional Examples

For a particle in one dimension, the Hamiltonian is:

$$\hat{H} = \frac{\hat{p}^2}{2m} + V(x), \quad \hat{p} = -i\hbar\frac{d}{dx}$$

and substituting this into the general Schrödinger equation gives:

$$-\frac{\hbar^2}{2m}\frac{d^2}{dx^2}\psi(x) + V(x)\psi(x) = E\psi(x)$$

This is the only case the Schrödinger equation is an ordinary differential equation, rather than a partial differential equation. The general solutions are always of the form:

$$\Psi(x,t) = \psi(x)e^{-iEt/\hbar}.$$

For N particles in one dimension, the Hamiltonian is:

$$\hat{H} = \sum_{n=1}^{N}\frac{\hat{p}_n^2}{2m_n} + V(x_1, x_2, \cdots x_N), \quad \hat{p}_n = -i\hbar\frac{\partial}{\partial x_n}$$

where the position of particle n is x_n. The corresponding Schrödinger equation is:

$$-\frac{\hbar^2}{2}\sum_{n=1}^{N}\frac{1}{m_n}\frac{\partial^2}{\partial x_n^2}\psi(x_1, x_2, \cdots x_N) + V(x_1, x_2, \cdots x_N)\psi(x_1, x_2, \cdots x_N) = E\psi(x_1, x_2, \cdots x_N).$$

so the general solutions have the form:

$$\Psi(x_1, x_2, \cdots x_N, t) = e^{-iEt/\hbar}\psi(x_1, x_2 \cdots x_N)$$

For non-interacting distinguishable particles, the potential of the system only influences each particle separately, so the total potential energy is the sum of potential energies for each particle:

$$V(x_1, x_2, \cdots x_N) = \sum_{n=1}^{N} V(x_n).$$

and the wavefunction can be written as a product of the wavefunctions for each particle:

$$\Psi(x_1, x_2, \cdots x_N, t) = e^{-iEt/\hbar}\prod_{n=1}^{N}\psi(x_n),$$

For non-interacting identical particles, the potential is still a sum, but wavefunction is a bit more complicated – it is a sum over the permutations of products of the separate wavefunctions to account for particle exchange. In general for interacting particles, the above decompositions are *not* possible.

Free Particle

For no potential, $V = 0$, so the particle is free and the equation reads:

$$-E\psi = \frac{\hbar^2}{2m}\frac{d^2\psi}{dx^2}$$

which has oscillatory solutions for $E > 0$ (the C_n are arbitrary constants):

$$\psi_E(x) = C_1 e^{i\sqrt{2mE/\hbar^2}\,x} + C_2 e^{-i\sqrt{2mE/\hbar^2}\,x}$$

and exponential solutions for $E < 0$

$$\psi_{-|E|}(x) = C_1 e^{\sqrt{2m|E|/\hbar^2}\,x} + C_2 e^{-\sqrt{2m|E|/\hbar^2}\,x}.$$

The exponentially growing solutions have an infinite norm, and are not physical. They are not allowed in a finite volume with periodic or fixed boundary conditions.

Constant Potential

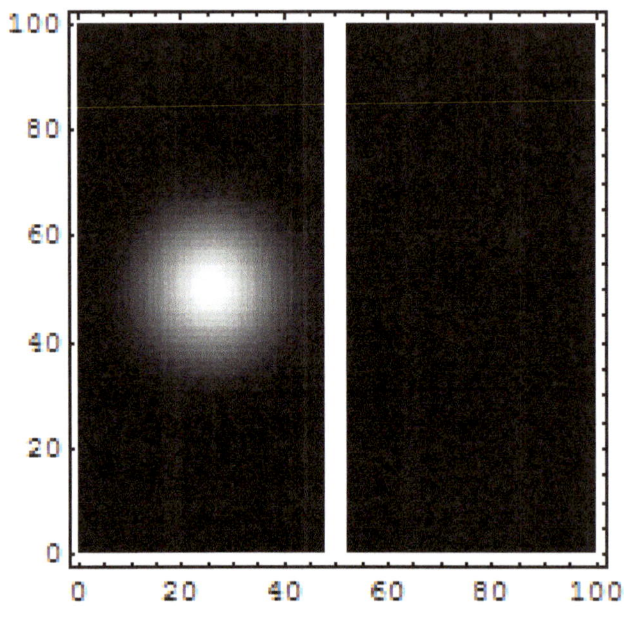

For a constant potential, $V = V_0$, the solution is oscillatory for $E > V_0$ and exponential for $E < V_0$, corresponding to energies that are allowed or disallowed in classical mechanics. Oscillatory solutions have a classically allowed energy and correspond to actual classical motions, while the exponential solutions have a disallowed energy and describe a small amount of quantum bleeding into the classically disallowed region, due to quantum tunneling. If the potential V_0 grows to infinity, the motion is classically confined to a finite region. Viewed far enough away, every solution is reduced to an exponential; the condition that the exponential is decreasing restricts the energy levels to a discrete set, called the allowed energies.

Harmonic Oscillator

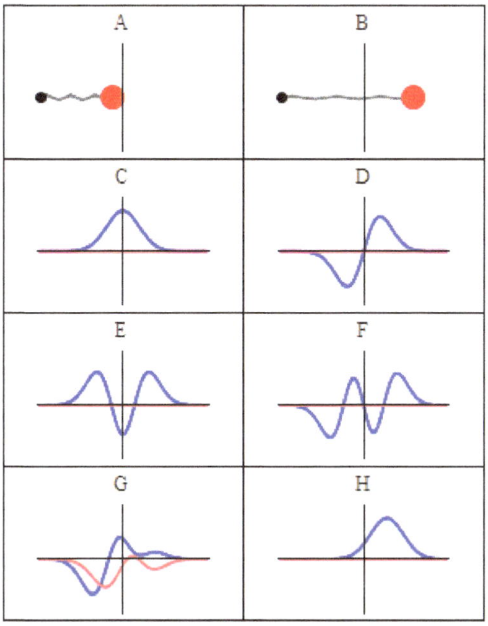

A harmonic oscillator in classical mechanics (A–B) and quantum mechanics (C–H). In (A–B), a ball, attached to a spring, oscillates back and forth. (C–H) are six solutions to the Schrödinger Equation for this situation. The horizontal axis is position, the vertical axis is the real part (blue) or imaginary part (red) of the wavefunction. Stationary states, or energy eigenstates, which are solutions to the time-independent Schrödinger equation, are shown in C, D, E, F, but not G or H.

The Schrödinger equation for this situation is

$$E\psi = -\frac{\hbar^2}{2m}\frac{d^2}{dx^2}\psi + \frac{1}{2}m\omega^2x^2\psi$$

It is a notable quantum system to solve for; since the solutions are exact (but complicated – in terms of Hermite polynomials), and it can describe or at least approximate a wide variety of other systems, including vibrating atoms, molecules, and atoms or ions in lattices, and approximating other potentials near equilibrium points. It is also the basis of perturbation methods in quantum mechanics.

There is a family of solutions – in the position basis they are

$$\psi_n(x) = \sqrt{\frac{1}{2^n n!}} \cdot \left(\frac{m\omega}{\pi\hbar}\right)^{1/4} \cdot e^{-\frac{m\omega x^2}{2\hbar}} \cdot H_n\left(\sqrt{\frac{m\omega}{\hbar}}x\right)$$

where $n = 0,1,2,\ldots$, and the functions H_n are the Hermite polynomials.

Three-dimensional Examples

The extension from one dimension to three dimensions is straightforward, all position and momentum operators are replaced by their three-dimensional expressions and the partial derivative with respect to space is replaced by the gradient operator.

The Hamiltonian for one particle in three dimensions is:

$$\hat{H} = \frac{\hat{\mathbf{p}} \cdot \hat{\mathbf{p}}}{2m} + V(\mathbf{r}), \quad \hat{\mathbf{p}} = -i\hbar\nabla$$

generating the equation:

$$-\frac{\hbar^2}{2m}\nabla^2\psi(\mathbf{r}) + V(\mathbf{r})\psi(\mathbf{r}) = E\psi(\mathbf{r})$$

with stationary state solutions of the form:

$$\Psi(\mathbf{r},t) = \psi(\mathbf{r})e^{-iEt/\hbar}$$

where the position of the particle is $\mathbf{r}$. Two useful coordinate systems for solving the Schrödinger equation are Cartesian coordinates so that $\mathbf{r} = (x, y, z)$ and spherical polar coordinates so that $\mathbf{r} = (r, \theta, \varphi)$, although other orthogonal coordinates are useful for solving the equation for systems with certain geometric symmetries.

For N particles in three dimensions, the Hamiltonian is:

$$\hat{H} = \sum_{n=1}^{N} \frac{\hat{\mathbf{p}}_n \cdot \hat{\mathbf{p}}_n}{2m_n} + V(\mathbf{r}_1, \mathbf{r}_2, \cdots \mathbf{r}_N), \quad \hat{\mathbf{p}}_n = -i\hbar\nabla_n$$

where the position of particle n is $\mathbf{r}_n$ and the gradient operators are partial derivatives with respect to the particle's position coordinates. In Cartesian coordinates, for particle n, the position vector is $\mathbf{r}_n = (x_n, y_n, z_n)$ while the gradient and Laplacian operator are respectively:

$$\nabla_n = \mathbf{e}_x \frac{\partial}{\partial x_n} + \mathbf{e}_y \frac{\partial}{\partial y_n} + \mathbf{e}_z \frac{\partial}{\partial z_n}, \quad \nabla_n^2 = \nabla_n \cdot \nabla_n = \frac{\partial^2}{\partial x_n{}^2} + \frac{\partial^2}{\partial y_n{}^2} + \frac{\partial^2}{\partial z_n{}^2}$$

The Schrödinger equation is:

$$-\frac{\hbar^2}{2}\sum_{n=1}^{N}\frac{1}{m_n}\nabla_n^2\Psi(\mathbf{r}_1, \mathbf{r}_2, \cdots \mathbf{r}_N) + V(\mathbf{r}_1, \mathbf{r}_2, \cdots \mathbf{r}_N)\Psi(\mathbf{r}_1, \mathbf{r}_2, \cdots \mathbf{r}_N) = E\Psi(\mathbf{r}_1, \mathbf{r}_2, \cdots \mathbf{r}_N)$$

with stationary state solutions:

$$\Psi(\mathbf{r}_1, \mathbf{r}_2 \cdots \mathbf{r}_N, t) = e^{-iEt/\hbar}\psi(\mathbf{r}_1, \mathbf{r}_2 \cdots \mathbf{r}_N)$$

Again, for non-interacting distinguishable particles the potential is the sum of particle potentials

$$V(\mathbf{r}_1, \mathbf{r}_2, \cdots \mathbf{r}_N) = \sum_{n=1}^{N} V(\mathbf{r}_n)$$

and the wavefunction is a product of the particle wavefuntions

$$\Psi(\mathbf{r}_1, \mathbf{r}_2 \cdots \mathbf{r}_N, t) = e^{-iEt/\hbar} \prod_{n=1}^{N} \psi(\mathbf{r}_n).$$

For non-interacting identical particles, the potential is a sum but the wavefunction is a sum over permutations of products. The previous two equations do not apply to interacting particles.

Following are examples where exact solutions are known.

Hydrogen Atom

This form of the Schrödinger equation can be applied to the hydrogen atom:

$$E\psi = -\frac{\hbar^2}{2\mu}\nabla^2\psi - \frac{e^2}{4\pi\varepsilon_0 r}\psi$$

where e is the electron charge, $\mathbf{r}$ is the position of the electron ($r = |\mathbf{r}|$ is the magnitude of the position), the potential term is due to the Coulomb interaction, wherein ε_0 is the electric constant (permittivity of free space) and

$$\mu = \frac{m_e m_p}{m_e + m_p}$$

is the 2-body reduced mass of the hydrogen nucleus (just a proton) of mass m_p and the electron of mass m_e. The negative sign arises in the potential term since the proton and electron are oppositely charged. The reduced mass in place of the electron mass is used since the electron and proton together orbit each other about a common centre of mass, and constitute a two-body problem to solve. The motion of the electron is of principle interest here, so the equivalent one-body problem is the motion of the electron using the reduced mass.

The wavefunction for hydrogen is a function of the electron's coordinates, and in fact can be separated into functions of each coordinate. Usually this is done in spherical polar coordinates:

$$\psi(r,\theta,\phi) = R(r)Y_\ell^m(\theta,\phi) = R(r)\Theta(\theta)\Phi(\phi)$$

where R are radial functions and $Y^m_\ell(\theta, \varphi)$ are spherical harmonics of degree ℓ and order m. This is the only atom for which the Schrödinger equation has been solved for exactly. Multi-electron atoms require approximative methods. The family of solutions are:

$$\psi_{n\ell m}(r,\theta,\phi) = \sqrt{\left(\frac{2}{na_0}\right)^3 \frac{(n-\ell-1)!}{2n[(n+\ell)!]}} e^{-r/na_0} \left(\frac{2r}{na_0}\right)^\ell L_{n-\ell-1}^{2\ell+1}\left(\frac{2r}{na_0}\right) \cdot Y_\ell^m(\theta,\phi)$$

where:

- $a_0 = \frac{4\pi\varepsilon_0\hbar^2}{m_e e^2}$ is the Bohr radius,
- $L_{n-\ell-1}^{2\ell+1}(\cdots)$ are the generalized Laguerre polynomials of degree $n - \ell - 1$.

- n, ℓ, m are the principal, azimuthal, and magnetic quantum numbers respectively: which take the values:

$$n = 1, 2, 3, \ldots$$
$$\ell = 0, 1, 2, \ldots, n-1$$
$$m = -\ell, \ldots, \ell$$

Two-electron Atoms or Ions

The equation for any two-electron system, such as the neutral helium atom (He, $Z = 2$), the negative hydrogen ion (H^-, $Z = 1$), or the positive lithium ion (Li^+, $Z = 3$) is:

$$E\psi = -\hbar^2\left[\frac{1}{2\mu}\left(\nabla_1^2 + \nabla_2^2\right) + \frac{1}{M}\nabla_1 \cdot \nabla_2\right]\psi + \frac{e^2}{4\pi\varepsilon_0}\left[\frac{1}{r_{12}} - Z\left(\frac{1}{r_1} + \frac{1}{r_2}\right)\right]\psi$$

where $\mathbf{r}_1$ is the position of one electron ($r_1 = |\mathbf{r}_1|$ is its magnitude), $\mathbf{r}_2$ is the position of the other electron ($r_2 = |\mathbf{r}_2|$ is the magnitude), $r_{12} = |\mathbf{r}_{12}|$ is the magnitude of the separation between them given by

$$|\mathbf{r}_{12}| = |\mathbf{r}_2 - \mathbf{r}_1|$$

μ is again the two-body reduced mass of an electron with respect to the nucleus of mass M, so this time

$$\mu = \frac{m_e M}{m_e + M}$$

and Z is the atomic number for the element (not a quantum number).

The cross-term of two laplacians

$$\frac{1}{M}\nabla_1 \cdot \nabla_2$$

is known as the *mass polarization term*, which arises due to the motion of atomic nuclei. The wavefunction is a function of the two electron's positions:

$$\psi = \psi(\mathbf{r}_1, \mathbf{r}_2).$$

There is no closed form solution for this equation.

Time Dependent

This is the equation of motion for the quantum state. In the most general form, it is written:

$$i\hbar\frac{\partial}{\partial t}\Psi = \hat{H}\Psi.$$

and the solution, the wavefunction, is a function of all the particle coordinates of the system and time. Following are specific cases.

For one particle in one dimension, the Hamiltonian

$$\hat{H} = \frac{\hat{p}^2}{2m} + V(x,t), \quad \hat{p} = -i\hbar\frac{\partial}{\partial x}$$

generates the equation:

$$i\hbar\frac{\partial}{\partial t}\Psi(x,t) = -\frac{\hbar^2}{2m}\frac{\partial^2}{\partial x^2}\Psi(x,t) + V(x,t)\Psi(x,t)$$

For N particles in one dimension, the Hamiltonian is:

$$\hat{H} = \sum_{n=1}^{N}\frac{\hat{p}_n^2}{2m_n} + V(x_1, x_2, \cdots x_N, t), \quad \hat{p}_n = -i\hbar\frac{\partial}{\partial x_n}$$

where the position of particle n is x_n, generating the equation:

$$i\hbar\frac{\partial}{\partial t}\Psi(x_1, x_2 \cdots x_N, t) = -\frac{\hbar^2}{2}\sum_{n=1}^{N}\frac{1}{m_n}\frac{\partial^2}{\partial x_n^2}\Psi(x_1, x_2 \cdots x_N, t) + V(x_1, x_2 \cdots x_N, t)\Psi(x_1, x_2 \cdots x_N, t).$$

For one particle in three dimensions, the Hamiltonian is:

$$\hat{H} = \frac{\hat{\mathbf{p}}\cdot\hat{\mathbf{p}}}{2m} + V(\mathbf{r},t), \quad \hat{\mathbf{p}} = -i\hbar\nabla$$

generating the equation:

$$i\hbar\frac{\partial}{\partial t}\Psi(\mathbf{r},t) = -\frac{\hbar^2}{2m}\nabla^2\Psi(\mathbf{r},t) + V(\mathbf{r},t)\Psi(\mathbf{r},t)$$

For N particles in three dimensions, the Hamiltonian is:

$$\hat{H} = \sum_{n=1}^{N}\frac{\hat{\mathbf{p}}_n\cdot\hat{\mathbf{p}}_n}{2m_n} + V(\mathbf{r}_1, \mathbf{r}_2, \cdots \mathbf{r}_N, t), \quad \hat{\mathbf{p}}_n = -i\hbar\nabla_n$$

where the position of particle n is $\mathbf{r}_n$, generating the equation:

$$i\hbar\frac{\partial}{\partial t}\Psi(\mathbf{r}_1, \mathbf{r}_2, \cdots \mathbf{r}_N, t) = -\frac{\hbar^2}{2}\sum_{n=1}^{N}\frac{1}{m_n}\nabla_n^2\Psi(\mathbf{r}_1, \mathbf{r}_2, \cdots \mathbf{r}_N, t) + V(\mathbf{r}_1, \mathbf{r}_2, \cdots \mathbf{r}_N, t)\Psi(\mathbf{r}_1, \mathbf{r}_2, \cdots \mathbf{r}_N, t)$$

This last equation is in a very high dimension, so the solutions are not easy to visualize.

Solution Methods

General techniques:	Methods for special cases:
• Perturbation theory • The variational method • Quantum Monte Carlo methods • Density functional theory • The WKB approximation and semi-classical expansion	• List of quantum-mechanical systems with analytical solutions • Hartree–Fock method and post Hartree–Fock methods

Properties

The Schrödinger equation has the following properties: some are useful, but there are shortcomings. Ultimately, these properties arise from the Hamiltonian used, and the solutions to the equation.

Linearity

In the development above, the Schrödinger equation was made to be linear for generality, though this has other implications. If two wave functions ψ_1 and ψ_2 are solutions, then so is any linear combination of the two:

$$\psi = a\psi_1 + b\psi_2$$

where a and b are any complex numbers (the sum can be extended for any number of wavefunctions). This property allows superpositions of quantum states to be solutions of the Schrödinger equation. Even more generally, it holds that a general solution to the Schrödinger equation can be found by taking a weighted sum over all single state solutions achievable. For example, consider a wave function $\Psi(x, t)$ such that the wave function is a product of two functions: one time independent, and one time dependent. If states of definite energy found using the time independent Schrödinger equation are given by $\psi_E(x)$ with amplitude A_n and time dependent phase factor is given by

$$e^{-iE_n t/\hbar},$$

then a valid general solution is

$$\Psi(x,t) = \sum_n A_n \psi_{E_n}(x) e^{-iE_n t/\hbar}.$$

Additionally, the ability to scale solutions allows one to solve for a wave function without normalizing it first. If one has a set of normalized solutions ψ_n, then

$$\Psi = \sum_n A_n \psi_n$$

can be normalized by ensuring that

$$\sum_n |A_n|^2 = 1.$$

This is much more convenient than having to verify that

$$\int_{-\infty}^{\infty} |\Psi(x)|^2 dx = \int_{-\infty}^{\infty} \Psi(x)\Psi^*(x) dx = 1.$$

Real Energy Eigenstates

For the time-independent equation, an additional feature of linearity follows: if two wave functions ψ_1 and ψ_2 are solutions to the time-independent equation with the same energy E, then so is any linear combination:

$$\hat{H}(a\psi_1 + b\psi_2) = a\hat{H}\psi_1 + b\hat{H}\psi_2 = E(a\psi_1 + b\psi_2).$$

Two different solutions with the same energy are called *degenerate*.

In an arbitrary potential, if a wave function ψ solves the time-independent equation, so does its complex conjugate, denoted ψ^*. By taking linear combinations, the real and imaginary parts of ψ are each solutions. If there is no degeneracy they can only differ by a factor.

In the time-dependent equation, complex conjugate waves move in opposite directions. If $\Psi(x, t)$ is one solution, then so is $\Psi^*(x, -t)$. The symmetry of complex conjugation is called time-reversal symmetry.

Space and Time Derivatives

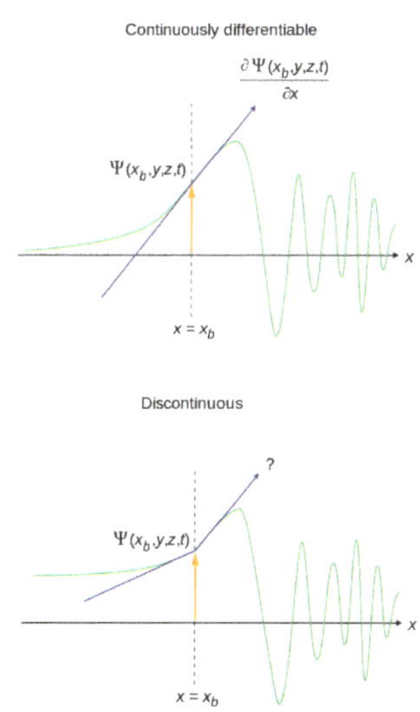

Continuity of the wavefunction and its first spatial derivative
(in the x direction, y and z coordinates not shown), at some time

The Schrödinger equation is first order in time and second in space, which describes the time evolution of a quantum state (meaning it determines the future amplitude from the present).

Explicitly for one particle in 3-dimensional Cartesian coordinates – the equation is

$$i\hbar\frac{\partial\Psi}{\partial t}=-\frac{\hbar^2}{2m}\left(\frac{\partial^2\Psi}{\partial x^2}+\frac{\partial^2\Psi}{\partial y^2}+\frac{\partial^2\Psi}{\partial z^2}\right)+V(x,y,z,t)\Psi.$$

The first time partial derivative implies the initial value (at t = 0) of the wavefunction

$$\Psi(x,y,z,0)$$

is an arbitrary constant. Likewise – the second order derivatives with respect to space implies the wavefunction *and* its first order spatial derivatives

$$\Psi(x_b,y_b,z_b,t)$$

$$\frac{\partial}{\partial x}\Psi(x_b,y_b,z_b,t)\quad\frac{\partial}{\partial y}\Psi(x_b,y_b,z_b,t)\quad\frac{\partial}{\partial z}\Psi(x_b,y_b,z_b,t)$$

are all arbitrary constants at a given set of points, where x_b, y_b, z_b are a set of points describing boundary b (derivatives are evaluated at the boundaries). Typically there are one or two boundaries, such as the step potential and particle in a box respectively.

As the first order derivatives are arbitrary, the wavefunction can be a continuously differentiable function of space, since at any boundary the gradient of the wavefunction can be matched.

On the contrary, wave equations in physics are usually *second order in time*, notable are the family of classical wave equations and the quantum Klein–Gordon equation.

Local Conservation of Probability

The Schrödinger equation is consistent with probability conservation. Multiplying the Schrödinger equation on the right by the complex conjugate wavefunction, and multiplying the wavefunction to the left of the complex conjugate of the Schrödinger equation, and subtracting, gives the continuity equation for probability:

$$\frac{\partial}{\partial t}\rho(\mathbf{r},t)+\nabla\cdot\mathbf{j}=0,$$

where

$$\rho=|\Psi|^2=\Psi^*(\mathbf{r},t)\Psi(\mathbf{r},t)$$

is the probability density (probability per unit volume, * denotes complex conjugate), and

$$\mathbf{j}=\frac{1}{2m}\left(\Psi^*\hat{\mathbf{p}}\Psi-\Psi\hat{\mathbf{p}}\Psi^*\right)$$

is the probability current (flow per unit area).

Hence predictions from the Schrödinger equation do not violate probability conservation.

Positive Energy

If the potential is bounded from below, meaning there is a minimum value of potential energy, the eigenfunctions of the Schrödinger equation have energy which is also bounded from below. This can be seen most easily by using the variational principle, as follows.

For any linear operator $\hat{A}$ bounded from below, the eigenvector with the smallest eigenvalue is the vector ψ that minimizes the quantity

$$\langle \psi | \hat{A} | \psi \rangle$$

over all ψ which are normalized. In this way, the smallest eigenvalue is expressed through the variational principle. For the Schrödinger Hamiltonian $\hat{H}$ bounded from below, the smallest eigenvalue is called the ground state energy. That energy is the minimum value of

$$\langle \psi | \hat{H} | \psi \rangle = \int \psi^*(\mathbf{r}) \left[-\frac{\hbar^2}{2m} \nabla^2 \psi(\mathbf{r}) + V(\mathbf{r})\psi(\mathbf{r}) \right] d^3\mathbf{r} = \int \left[\frac{\hbar^2}{2m} |\nabla \psi|^2 + V(\mathbf{r}) |\psi|^2 \right] d^3\mathbf{r} = \langle \hat{H} \rangle$$

(using integration by parts). Due to the complex modulus of ψ^2 (which is positive definite), the right hand side always greater than the lowest value of $V(x)$. In particular, the ground state energy is positive when $V(x)$ is everywhere positive.

For potentials which are bounded below and are not infinite over a region, there is a ground state which minimizes the integral above. This lowest energy wavefunction is real and positive definite – meaning the wavefunction can increase and decrease, but is positive for all positions. It physically cannot be negative: if it were, smoothing out the bends at the sign change (to minimize the wavefunction) rapidly reduces the gradient contribution to the integral and hence the kinetic energy, while the potential energy changes linearly and less quickly. The kinetic and potential energy are both changing at different rates, so the total energy is not constant, which can't happen (conservation). The solutions are consistent with Schrödinger equation if this wavefunction is positive definite.

The lack of sign changes also shows that the ground state is nondegenerate, since if there were two ground states with common energy E, not proportional to each other, there would be a linear combination of the two that would also be a ground state resulting in a zero solution.

Analytic Continuation to Diffusion

The above properties (positive definiteness of energy) allow the analytic continuation of the Schrödinger equation to be identified as a stochastic process. This can be interpreted as the Huygens–Fresnel principle applied to De Broglie waves; the spreading wavefronts are diffusive probability amplitudes. For a free particle (not subject to a potential) in a random walk, substituting $\tau = it$ into the time-dependent Schrödinger equation gives:

$$\frac{\partial}{\partial \tau} X(\mathbf{r}, \tau) = \frac{\hbar}{2m} \nabla^2 X(\mathbf{r}, \tau), \quad X(\mathbf{r}, \tau) = \Psi(\mathbf{r}, \tau / i)$$

which has the same form as the diffusion equation, with diffusion coefficient $\frac{\hbar}{2m}$. In that case, the diffusivity yields the De Broglie relation in accordance with the Markov process.

Regularity

On the space L^2 of square-integrable densities, the Schrödinger semigroup $e^{it\hat{H}}$ is a unitary evolution, and therefore surjective. The flows satisfy the Schrödinger equation $i\partial_t u = \hat{H}u$, where the derivative is taken in the distribution sense. However, since $\hat{H}$ for most physically reasonable Hamiltonians (e.g., the Laplace operator, possibly modified by a potential) is unbounded in L^2, this shows that the semigroup flows lack Sobolev regularity in general. Instead, solutions of the Schrödinger equation satisfy a Strichartz estimate.

Relativistic Quantum Mechanics

Relativistic quantum mechanics is obtained where quantum mechanics and special relativity simultaneously apply. In general, one wishes to build relativistic wave equations from the relativistic energy–momentum relation

$$E^2 = (pc)^2 + (m_0 c^2)^2$$

instead of classical energy equations. The Klein–Gordon equation and the Dirac equation are two such equations. The Klein–Gordon equation,

$$\frac{1}{c^2}\frac{\partial^2}{\partial t^2}\psi - \nabla^2\psi + \frac{m^2c^2}{\hbar^2}\psi = 0.$$

was the first such equation to be obtained, even before the nonrelativistic one, and applies to massive spinless particles. The Dirac equation arose from taking the "square root" of the Klein–Gordon equation by factorizing the entire relativistic wave operator into a product of two operators – one of these is the operator for the entire Dirac equation.

The general form of the Schrödinger equation remains true in relativity, but the Hamiltonian is less obvious. For example, the Dirac Hamiltonian for a particle of mass m and electric charge q in an electromagnetic field (described by the electromagnetic potentials φ and $\mathbf{A}$) is:

$$\hat{H}_{\text{Dirac}} = \gamma^0\left[c\gamma\cdot\left(\hat{\mathbf{p}} - q\mathbf{A}\right) + mc^2 + \gamma^0 q\phi\right],$$

in which the $\gamma = (\gamma^1, \gamma^2, \gamma^3)$ and γ^0 are the Dirac gamma matrices related to the spin of the particle. The Dirac equation is true for all spin-$\frac{1}{2}$ particles, and the solutions to the equation are 4-component spinor fields with two components corresponding to the particle and the other two for the antiparticle.

For the Klein–Gordon equation, the general form of the Schrödinger equation is inconvenient to use, and in practice the Hamiltonian is not expressed in an analogous way to the Dirac Hamiltonian. The equations for relativistic quantum fields can be obtained in other ways, such as starting from a Lagrangian density and using the Euler–Lagrange equations for fields, or use the representation theory of the Lorentz group in which certain representations can be used to fix the equation for a free particle of given spin (and mass).

In general, the Hamiltonian to be substituted in the general Schrödinger equation is not just a function of the position and momentum operators (and possibly time), but also of spin matrices. Also, the solutions to a relativistic wave equation, for a massive particle of spin s, are complex-valued $2(2s + 1)$-component spinor fields.

Quantum Field Theory

The general equation is also valid and used in quantum field theory, both in relativistic and non-relativistic situations. However, the solution ψ is no longer interpreted as a "wave", but should be interpreted as an operator acting on states existing in a Fock space.

First Order Form

The Schrödinger equation can also be derived from a first order form similar to the manner in which the Klein-Gordon equation can be derived from the Dirac equation. In 1D the first order equation is given by

$$-i\partial_z \psi = (i\eta\partial_t + \eta^\dagger m)\psi$$

This equation allows for the inclusion of spin in nonrelativistic quantum mechanics. Squaring the above equation yields the Schrödinger equation in 1D. The matrices η obey the following properties

$$\eta^2 = 0$$
$$(\eta^\dagger)^2 = 0$$
$$\left\{\eta, \eta^\dagger\right\} = 2I$$

The 3 dimensional version of the equation is given by

$$-i\gamma_i \partial_i \psi = (i\eta\partial_t + \eta^\dagger m)\psi$$

Here $\eta = (\gamma_0 + i\gamma_5)/\sqrt{2}$ is a 4×4 nilpotent matrix and γ_i are the Dirac gamma matrices ($i = 1,2,3$). The Schrödinger equation in 3D can be obtained by squaring the above equation. In the nonrelativistic limit $E - m \simeq E'$ and $E + m \simeq 2m$, , the above equation can be derived from the Dirac equation.

In 1926, Schrodinger said,

There is no reason to retain even Bohr's electron orbits,

Basic equation governs wave motions and electron in a nuclear field.

If we consider, the electron as wave defined by de Broglie, we can treat the wave equation.

One dimensional wave propagation freely in space along x-coordinate

$$\nabla^2 \psi(x,t) = \frac{1}{c^2}\frac{\partial^2 \psi(x,t)}{\partial t^2}$$

$\psi(x,t)$ represents the wave nature of the electron, WAVEFUNCTION

Since we are looking for a solution of stationary states, we separate the space (x) and time (t) part of the wavefunction such as

$$\psi(x,t) = \phi(x)e^{2\pi i v t}$$

Now,

$$\frac{\partial^2 \psi}{\partial t^2} = -4\pi^2 v^2 \phi(x) e^{2\pi i v t} \text{ and } \frac{\partial^2 \psi}{\partial x^2} = e^{2\pi i v t}\frac{\partial^2 \phi}{\partial x^2}$$

Substituting the equation above in the wave equation, we get

$$\begin{aligned} \frac{d^2\phi(x)}{dx^2} &= -\frac{4\pi^2 v^2}{c^2}\phi(x) \\ &= -\frac{4\pi^2 p^2 \not{c}^2}{\not{c}^2 h^2}\phi(x) \\ &= -\frac{p^2}{\hbar^2}\phi(x) \qquad \left[\because \hbar = \frac{h}{2\pi}\right] \end{aligned}$$

According to de Broglie wavelength,

$$\because v = \frac{c}{\lambda} = \frac{c}{h/p} = \frac{pc}{h} \;\&\; p = \frac{h}{\lambda}$$

And the total energy (E) = kinetic energy + potential energy (V)

$$\because \frac{p^2}{2m} + V = E$$
$$\Rightarrow p^2 = 2m(E - V)$$

Substituting the equation above, we get

$$\frac{d^2\phi(x)}{dx^2} = -\frac{2m}{\hbar^2}(E - V)\phi(x)$$
$$\Rightarrow -\frac{\hbar}{2m}\frac{d^2\phi(x)}{dx^2} = (E - V)\phi(x)$$
$$\Rightarrow -\frac{\hbar^2}{2m}\frac{d^2\phi(x)}{dx^2} + V\phi(x) = E\phi(x)$$

This is known as one dimension time independent Schrodinger Equation

For the quantum mechanical treatment, we will convert the classical dynamical quantities ($px, py,$ and pz) to its corresponding quantum mechanical operators.

$$p_x \rightarrow -i\hbar\frac{\partial}{\partial x}, p_y \rightarrow -i\hbar\frac{\partial}{\partial y}, \rightarrow -i\hbar\frac{\partial}{\partial z}$$

$$E \rightarrow i\hbar\frac{\partial}{\partial t}$$

Substituting the equation above, we get

$$-\frac{\hbar}{2\mu}\left(\frac{\partial^2}{\partial x^2}+\frac{\partial^2}{\partial y^2}+\frac{\partial^2}{\partial z^2}\right)+V(x,y,z)=i\hbar\frac{\partial}{\partial t}$$

Here, we introduce the Wavefuntion to represent the electron.

$$\Psi = \Psi(x,y,z,t)$$

The wavefunction contains the information about the position and the time evolution of the electronic motion. We will understand more about this later.

So operating the equation abvoe, on the wavefunction,

$$\Psi = \Psi(x,y,z,t)$$

$$-\frac{\hbar^3}{2\mu}\left(\frac{\partial^3\Psi(x,y,z,t)}{\partial x^3}+\frac{\partial^3\Psi(x,y,z,t)}{\partial y^3}+\frac{\partial^3\Psi(x,y,z,t)}{\partial z^3}\right)+V(x,y,z)\Psi(x,y,z,t)=i\hbar\frac{\partial\Psi(x,y,z,t)}{\partial t}$$

$$\Rightarrow -\frac{\hbar^3}{2\mu}\nabla^3\Psi+V\Psi=i\hbar\frac{\partial\Psi}{\partial t}$$

This Equation is known as Time dependent Schrödinger Equation, where

$$\nabla^2 = \frac{\partial^2}{\partial x^2}+\frac{\partial^2}{\partial y^2}+\frac{\partial^2}{\partial z^2} \quad \textit{is the Laplacian operator.}$$

The three dimensional Schrodinger Equation is

$$-\frac{\hbar^2}{2m}\frac{d^2\phi(r)}{dx^2}+V\phi(r)=E\phi(r)$$

If we take the potential of the electron due to nucleus is

$$V = -\frac{Ze^2}{r}$$

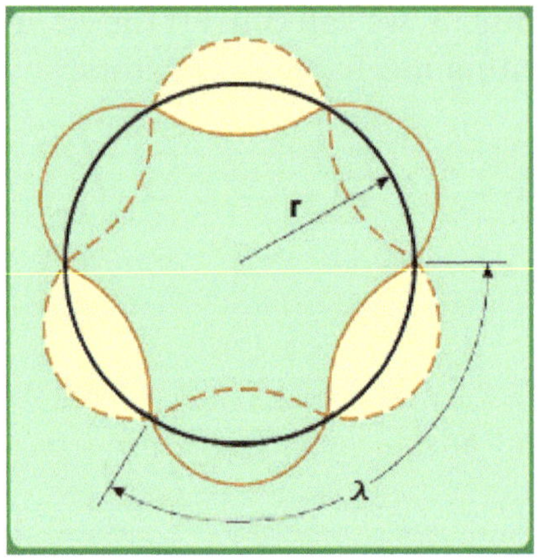

Where r is the distance between the electron and the nucleus and Z is the charge (Z = 1 for hydrogen).

Substituting the potential in the Schrödinger Equation, the energy becomes

$$E_n = -\frac{me^4 z^2}{2\hbar^2}\frac{1}{n^2}$$

Where n is known as the principal quantum number. Details of this quantum mechanical calculation are given later.

This energy is exactly the same energy expression derived by Bohr. In this figure, E (n=1) = -110000 cm^{-1}.

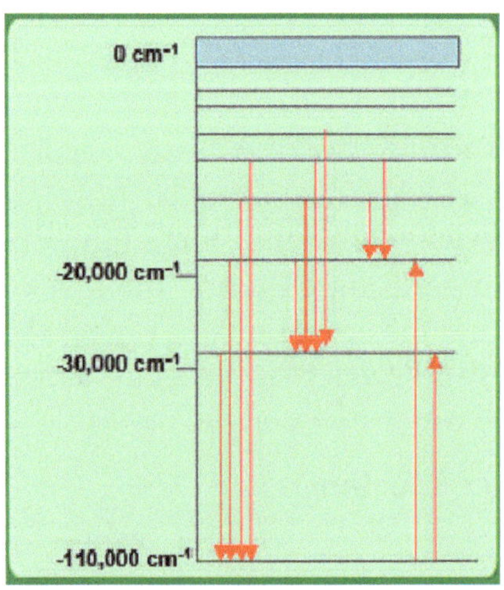

Wave–particle Duality

Wave–particle duality is the concept that every elementary particle or quantic entity may be partly described in terms not only of particles, but also of waves. It expresses the inability of the classical concepts "particle" or "wave" to fully describe the behavior of quantum-scale objects. As Albert Einstein wrote: "*It seems as though we must use sometimes the one theory and sometimes the other, while at times we may use either. We are faced with a new kind of difficulty. We have two contradictory pictures of reality; separately neither of them fully explains the phenomena of light, but together they do.*"

Through the work of Max Planck, Einstein, Louis de Broglie, Arthur Compton, Niels Bohr and many others, current scientific theory holds that all particles also have a wave nature (and vice versa). This phenomenon has been verified not only for elementary particles, but also for compound particles like atoms and even molecules. For macroscopic particles, because of their extremely short wavelengths, wave properties usually cannot be detected.

Although the use of the wave-particle duality has worked well in physics, the *meaning* or *interpretation* has not been satisfactorily resolved.

Niels Bohr regarded the "duality paradox" as a fundamental or metaphysical fact of nature. A given kind of quantum object will exhibit sometimes wave, sometimes particle, character, in respectively different physical settings. He saw such duality as one aspect of the concept of complementarity. Bohr regarded renunciation of the cause-effect relation, or complementarity, of the space-time picture, as essential to the quantum mechanical account.

Werner Heisenberg considered the question further. He saw the duality as present for all quantic entities, but not quite in the usual quantum mechanical account considered by Bohr. He saw it in what is called second quantization, which generates an entirely new concept of fields which exist in ordinary space-time, causality still being visualizable. Classical field values (e.g. the electric and magnetic field strengths of Maxwell) are replaced by an entirely new kind of field value, as considered in quantum field theory. Turning the reasoning around, ordinary quantum mechanics can be deduced as a specialized consequence of quantum field theory.

Brief History of Wave and Particle Viewpoints

Democritus—the original *atomist*—argued that all things in the universe, including light, are composed of indivisible sub-components (light being some form of solar atom). At the beginning of the 11th Century, the Arabic scientist Alhazen wrote the first comprehensive treatise on optics; describing refraction, reflection, and the operation of a pinhole lens via rays of light traveling from the point of emission to the eye. He asserted that these rays were composed of particles of light. In 1630, René Descartes popularized and accredited the opposing wave description in his treatise on light, showing that the behavior of light could be re-created by modeling wave-like disturbances in a universal medium ("plenum"). Beginning in 1670 and progressing over three decades, Isaac Newton developed and championed his corpuscular hypothesis, arguing that the perfectly straight lines of reflection demonstrated light's particle nature; only particles could travel in such straight lines. He explained refraction by positing that particles of light accelerated laterally upon entering a denser medium. Around the same time, Newton's contemporaries Robert Hooke and Christiaan Huygens—and later Augustin-Jean Fresnel—mathematically refined the wave viewpoint, showing that if light traveled at different speeds in different media (such as water and air), refraction could be easily explained as the medium-dependent propagation of light waves. The resulting Huygens–Fresnel principle was extremely successful at reproducing light's behavior and was subsequently supported by Thomas Young's 1803 discovery of double-slit interference. The wave view did not immediately displace the ray and particle view, but began to dominate scientific thinking about light in the mid 19th century, since it could explain polarization phenomena that the alternatives could not.

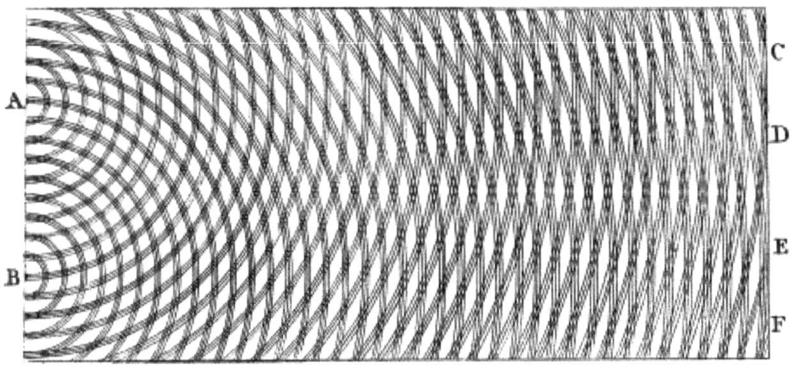

Thomas Young's sketch of two-slit diffraction of waves, 1803

James Clerk Maxwell discovered that he could apply his equations for electromagnetism, which had been previously discovered, along with a slight modification to describe self-propagating waves of oscillating electric and magnetic fields. When the propagation speed of these electromagnetic waves was calculated, the speed of light fell out. It quickly became apparent that visible light, ultraviolet light, and infrared light (phenomena thought previously to be unrelated) were all electromagnetic waves of differing frequency. The wave theory had prevailed—or at least it seemed to.

While the 19th century had seen the success of the wave theory at describing light, it had also witnessed the rise of the atomic theory at describing matter. Antoine Lavoisier deduced the law of conservation of mass and categorized many new chemical elements and compounds; and Joseph Louis Proust advanced chemistry towards the atom by showing that elements combined in definite proportions. This led John Dalton to propose that elements were invisible sub components; Amedeo Avogadro discovered diatomic gases and completed the basic atomic theory, allowing the correct molecular formulae of most known compounds—as well as the correct weights of atoms—to be deduced and categorized in a consistent manner. Dimitri Mendeleev saw an order in recurring chemical properties, and created a table presenting the elements in unprecedented order and symmetry.

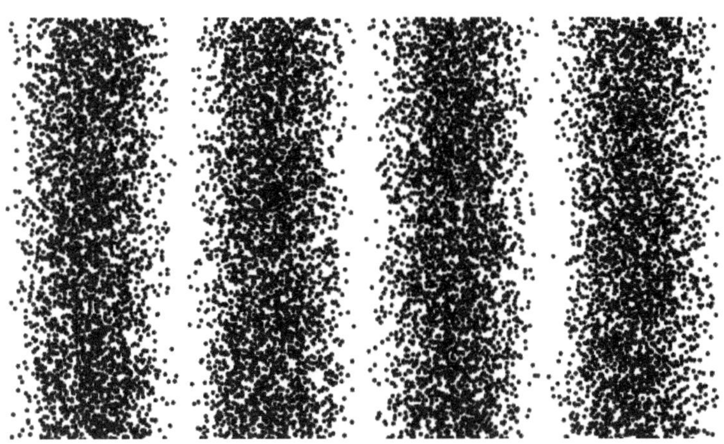

Particle impacts make visible the interference pattern of waves

A quantum particle is represented by a wave packet

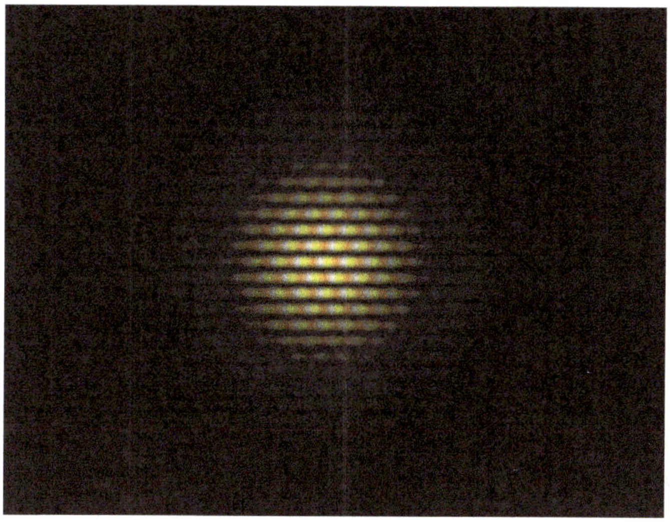

Interference of a quantum particle with itself

Turn of the 20th Century and the Paradigm Shift

Particles of Electricity

At the close of the 19th century, the reductionism of atomic theory began to advance into the atom itself; determining, through physics, the nature of the atom and the operation of chemical reactions. Electricity, first thought to be a fluid, was now understood to consist of particles called electrons. This was first demonstrated by J. J. Thomson in 1897 when, using a cathode ray tube, he found that an electrical charge would travel across a vacuum (which would possess infinite resistance in classical theory). Since the vacuum offered no medium for an electric fluid to travel, this discovery could only be explained via a particle carrying a negative charge and moving through the vacuum. This *electron* flew in the face of classical electrodynamics, which had successfully treated electricity as a fluid for many years (leading to the invention of batteries, electric motors, dynamos, and arc lamps). More importantly, the intimate relation between electric charge and electromagnetism had been well documented following the discoveries of Michael Faraday and James Clerk Maxwell. Since electromagnetism was *known* to be a wave generated by a changing electric or magnetic *field* (a continuous, wave-like entity itself) an atomic/particle description of electricity and charge was a non sequitur. Furthermore, classical electrodynamics was not the only classical theory rendered incomplete.

Radiation Quantization

In 1901, Max Planck published an analysis that succeeded in reproducing the observed spectrum of light emitted by a glowing object. To accomplish this, Planck had to make an ad hoc mathematical assumption of quantized energy of the oscillators (atoms of the black body) that emit radiation. Einstein later proposed that electromagnetic radiation itself is quantized, not the energy of radiating atoms.

Black-body radiation, the emission of electromagnetic energy due to an object's heat, could not be explained from classical arguments alone. The equipartition theorem of classical mechanics, the basis of all classical thermodynamic theories, stated that an object's energy is partitioned equally among the object's vibrational modes. But applying the same reasoning to the electromagnetic emission of such a thermal object was not so successful. That thermal objects emit light had been long known. Since light was known to be waves of electromagnetism, physicists hoped to describe this emission via classical laws. This became known as the black body problem. Since the equipartition theorem worked so well in describing the vibrational modes of the thermal object itself, it was natural to assume that it would perform equally well in describing the radiative emission of such objects. But a problem quickly arose: if each mode received an equal partition of energy, the short wavelength modes would consume all the energy. This became clear when plotting the Rayleigh–Jeans law which, while correctly predicting the intensity of long wavelength emissions, predicted infinite total energy as the intensity diverges to infinity for short wavelengths. This became known as the ultraviolet catastrophe.

In 1900, Max Planck hypothesized that the frequency of light emitted by the black body depended on the frequency of the *oscillator* that emitted it, and the energy of these oscillators increased linearly with frequency (according to his constant h, where $E = h\nu$). This was not an unsound proposal considering that macroscopic oscillators operate similarly: when studying five simple harmonic oscillators of equal amplitude but different frequency, the oscillator with the highest frequency possesses the highest energy (though this relationship is not linear like Planck's). By demanding that high-frequency light must be emitted by an oscillator of equal frequency, and further requiring that this oscillator occupy higher energy than one of a lesser frequency, Planck avoided any catastrophe; giving an equal partition to high-frequency oscillators produced successively fewer oscillators and less emitted light. And as in the Maxwell–Boltzmann distribution, the low-frequency, low-energy oscillators were suppressed by the onslaught of thermal jiggling from higher energy oscillators, which necessarily increased their energy and frequency.

The most revolutionary aspect of Planck's treatment of the black body is that it inherently relies on an integer number of oscillators in thermal equilibrium with the electromagnetic field. These oscillators *give* their entire energy to the electromagnetic field, creating a quantum of light, as often as they are *excited* by the electromagnetic field, absorbing a quantum of light and beginning to oscillate at the corresponding frequency. Planck had intentionally created an atomic theory of the black body, but had unintentionally generated an atomic theory of light, where the black body never generates quanta of light at a given frequency with an energy less than $\mathbf{h\nu}$. However, once realizing that he had quantized the electromagnetic field, he denounced particles of light as a limitation of his approximation, not a property of reality.

Photoelectric Effect Illuminated

While Planck had solved the ultraviolet catastrophe by using atoms and a quantized electromagnetic field, most contemporary physicists agreed that Planck's "light quanta" represented only flaws in his model. A more-complete derivation of black body radiation would yield a fully continuous and 'wave-like' electromagnetic field with no quantization. However, in 1905 Albert Einstein took Planck's black body model to produce his solution to another outstanding problem of the day: the photoelectric effect, wherein electrons are emitted from atoms when they absorb energy from light. Since their discovery eight years previously, electrons had been studied in physics laboratories worldwide.

In 1902 Philipp Lenard discovered that the energy of these ejected electrons did *not* depend on the intensity of the incoming light, but instead on its *frequency*. So if one shines a little low-frequency light upon a metal, a few low energy electrons are ejected. If one now shines a very intense beam of low-frequency light upon the same metal, a whole slew of electrons are ejected; however they possess the same low energy, there are merely *more of them*. The more light there is, the more electrons are ejected. Whereas in order to get high energy electrons, one must illuminate the metal with high-frequency light. Like blackbody radiation, this was at odds with a theory invoking continuous transfer of energy between radiation and matter. However, it can still be explained using a fully classical description of light, as long as matter is quantum mechanical in nature.

If one used Planck's energy quanta, and demanded that electromagnetic radiation at a given frequency could only transfer energy to matter in integer multiples of an energy quantum **hv**, then the photoelectric effect could be explained very simply. Low-frequency light only ejects low-energy electrons because each electron is excited by the absorption of a single photon. Increasing the intensity of the low-frequency light (increasing the number of photons) only increases the number of excited electrons, not their energy, because the energy of each photon remains low. Only by increasing the frequency of the light, and thus increasing the energy of the photons, can one eject electrons with higher energy. Thus, using Planck's constant h to determine the energy of the photons based upon their frequency, the energy of ejected electrons should also increase linearly with frequency; the gradient of the line being Planck's constant. These results were not confirmed until 1915, when Robert Andrews Millikan, who had previously determined the charge of the electron, produced experimental results in perfect accord with Einstein's predictions. While the energy of ejected electrons reflected Planck's constant, the existence of photons was not explicitly proven until the discovery of the photon antibunching effect, of which a modern experiment can be performed in undergraduate-level labs. This phenomenon could only be explained via photons, and not through any semi-classical theory (which could alternatively explain the photoelectric effect). When Einstein received his Nobel Prize in 1921, it was not for his more difficult and mathematically laborious special and general relativity, but for the simple, yet totally revolutionary, suggestion of quantized light. Einstein's "light quanta" would not be called photons until 1925, but even in 1905 they represented the quintessential example of wave-particle duality. Electromagnetic radiation propagates following linear wave equations, but can only be emitted or absorbed as discrete elements, thus acting as a wave and a particle simultaneously.

Einstein's Explanation of the Photoelectric Effect

In 1905, Albert Einstein provided an explanation of the photoelectric effect, a hitherto troubling experiment that the wave theory of light seemed incapable of explaining. He did so by postulating the existence of photons, quanta of light energy with particulate qualities.

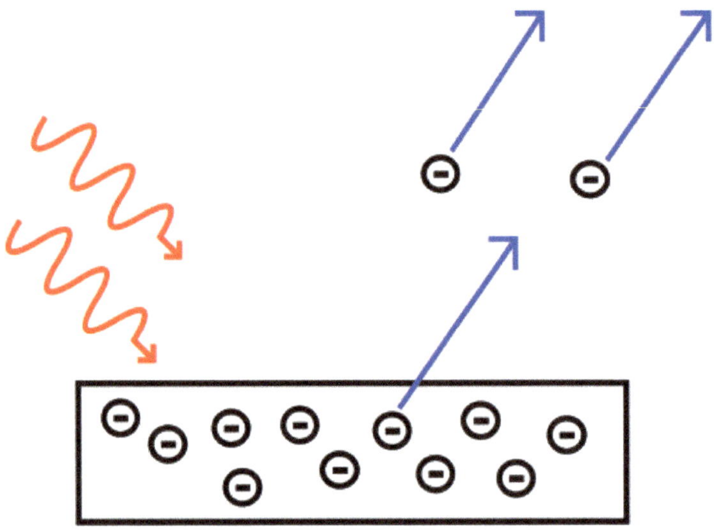

The photoelectric effect. Incoming photons on the left strike a metal plate (bottom), and eject electrons, depicted as flying off to the right

In the photoelectric effect, it was observed that shining a light on certain metals would lead to an electric current in a circuit. Presumably, the light was knocking electrons out of the metal, causing current to flow. However, using the case of potassium as an example, it was also observed that while a dim blue light was enough to cause a current, even the strongest, brightest red light available with the technology of the time caused no current at all. According to the classical theory of light and matter, the strength or amplitude of a light wave was in proportion to its brightness: a bright light should have been easily strong enough to create a large current. Yet, oddly, this was not so.

Einstein explained this conundrum by postulating that the electrons can receive energy from electromagnetic field only in discrete portions (quanta that were called photons): an amount of energy E that was related to the frequency f of the light by

$$E = hf$$

where h is Planck's constant (6.626×10^{-34} J seconds). Only photons of a high enough frequency (above a certain *threshold* value) could knock an electron free. For example, photons of blue light had sufficient energy to free an electron from the metal, but photons of red light did not. One photon of light above the threshold frequency could release only one electron; the higher the frequency of a photon, the higher the kinetic energy of the emitted electron, but no amount of light (using technology available at the time) below the threshold frequency could release an electron. To "violate" this law would require extremely high-intensity lasers which had not yet been invented. Intensity-dependent phenomena have now been studied in detail with such lasers.

Einstein was awarded the Nobel Prize in Physics in 1921 for his discovery of the law of the photoelectric effect.

De Broglie's Wavelength

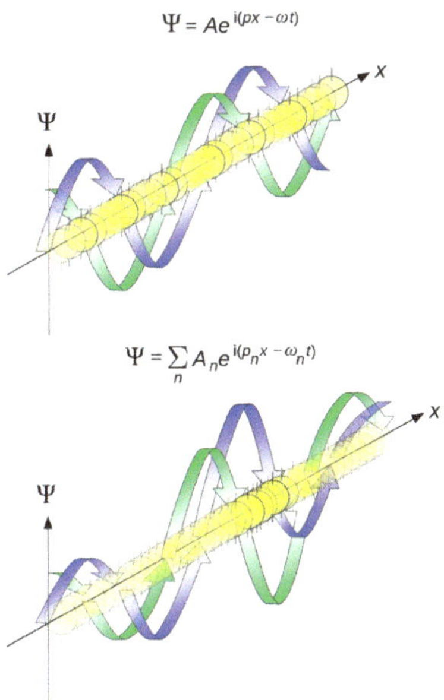

Propagation of de Broglie waves in 1d—real part of the complex amplitude is blue, imaginary part is green. The probability (shown as the colour opacity) of finding the particle at a given point x is spread out like a waveform; there is no definite position of the particle. As the amplitude increases above zero the curvature decreases, so the amplitude decreases again, and vice versa—the result is an alternating amplitude: a wave. Top: Plane wave. Bottom: Wave packet

In 1924, Louis-Victor de Broglie formulated the de Broglie hypothesis, claiming that *all* matter, not just light, has a wave-like nature; he related wavelength (denoted as λ), and momentum (denoted as p):

$$\lambda = \frac{h}{p}$$

This is a generalization of Einstein's equation above, since the momentum of a photon is given by $p = \frac{E}{c}$ and the wavelength (in a vacuum) by $\lambda = \frac{c}{f}$, where c is the speed of light in vacuum.

De Broglie's formula was confirmed three years later for electrons (which differ from photons in having a rest mass) with the observation of electron diffraction in two independent experiments. At the University of Aberdeen, George Paget Thomson passed a beam of electrons through a thin metal film and observed the predicted interference patterns. At Bell Labs, Clinton Joseph Davisson and Lester Halbert Germer guided their beam through a crystalline grid.

De Broglie was awarded the Nobel Prize for Physics in 1929 for his hypothesis. Thomson and Davisson shared the Nobel Prize for Physics in 1937 for their experimental work.

Heisenberg's Uncertainty Principle

In his work on formulating quantum mechanics, Werner Heisenberg postulated his uncertainty principle, which states:

$$\Delta x \Delta p \geq \frac{\hbar}{2}$$

where

> Δ here indicates standard deviation, a measure of spread or uncertainty;
>
> x and p are a particle's position and linear momentum respectively.
>
> $\hbar$ is the reduced Planck's constant (Planck's constant divided by 2π).

Heisenberg originally explained this as a consequence of the process of measuring: Measuring position accurately would disturb momentum and vice versa, offering an example (the "gamma-ray microscope") that depended crucially on the de Broglie hypothesis. The thought is now, however, that this only partly explains the phenomenon, but that the uncertainty also exists in the particle itself, even before the measurement is made.

In fact, the modern explanation of the uncertainty principle, extending the Copenhagen interpretation first put forward by Bohr and Heisenberg, depends even more centrally on the wave nature of a particle: Just as it is nonsensical to discuss the precise location of a wave on a string, particles do not have perfectly precise positions; likewise, just as it is nonsensical to discuss the wavelength of a "pulse" wave traveling down a string, particles do not have perfectly precise momenta (which corresponds to the inverse of wavelength). Moreover, when position is relatively well defined, the wave is pulse-like and has a very ill-defined wavelength (and thus momentum). And conversely, when momentum (and thus wavelength) is relatively well defined, the wave looks long and sinusoidal, and therefore it has a very ill-defined position.

De Broglie–Bohm Theory

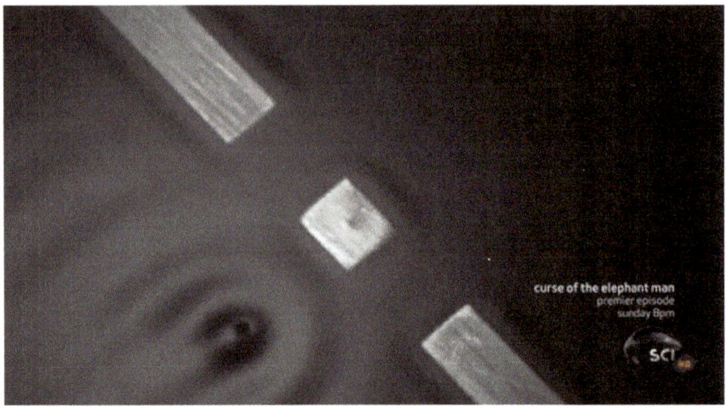

Couder experiments, "materializing" the *pilot wave* model

De Broglie himself had proposed a pilot wave construct to explain the observed wave-particle duality. In this view, each particle has a well-defined position and momentum, but is guided by a wave function derived from Schrödinger's equation. The pilot wave theory was initially rejected because it generated non-local effects when applied to systems involving more than one particle. Non-locality, however, soon became established as an integral feature of quantum theory, and David Bohm extended de Broglie's model to explicitly include it.

In the resulting representation, also called the de Broglie–Bohm theory or Bohmian mechanics, the wave-particle duality vanishes, and explains the wave behaviour as a scattering with wave appearance, because the particle's motion is subject to a guiding equation or quantum potential.

"This idea seems to me so natural and simple, to resolve the wave-particle dilemma in such a clear and ordinary way, that it is a great mystery to me that it was so generally ignored", J.S. Bell.

The best illustration of the *pilot-wave model* was given by Couder's 2010 "walking droplets" experiments, demonstrating the pilot-wave behaviour in a macroscopic mechanical analog.

Wave Behavior of Large Objects

Since the demonstrations of wave-like properties in photons and electrons, similar experiments have been conducted with neutrons and protons. Among the most famous experiments are those of Estermann and Otto Stern in 1929. Authors of similar recent experiments with atoms and molecules, described below, claim that these larger particles also act like waves. A wave is basically a group of particles which moves in a particular form of motion, i.e. to and fro. If we break that flow by an object it will convert into radiants.

A dramatic series of experiments emphasizing the action of gravity in relation to wave–particle duality was conducted in the 1970s using the neutron interferometer. Neutrons, one of the components of the atomic nucleus, provide much of the mass of a nucleus and thus of ordinary matter. In the neutron interferometer, they act as quantum-mechanical waves directly subject to the force of gravity. While the results were not surprising since gravity was known to act on everything, including light, the self-interference of the quantum mechanical wave of a massive fermion in a gravitational field had never been experimentally confirmed before.

In 1999, the diffraction of C_{60} fullerenes by researchers from the University of Vienna was reported. Fullerenes are comparatively large and massive objects, having an atomic mass of about 720 u. The de Broglie wavelength of the incident beam was about 2.5 pm, whereas the diameter of the molecule is about 1 nm, about 400 times larger. In 2012, these far-field diffraction experiments could be extended to phthalocyanine molecules and their heavier derivatives, which are composed of 58 and 114 atoms respectively. In these experiments the build-up of such interference patterns could be recorded in real time and with single molecule sensitivity.

In 2003, the Vienna group also demonstrated the wave nature of tetraphenylporphyrin—a flat biodye with an extension of about 2 nm and a mass of 614 u. For this demonstration they employed a near-field Talbot Lau interferometer. In the same interferometer they also found interference fringes for $C_{60}F_{48}$, a fluorinated buckyball with a mass of about 1600 u, composed of 108 atoms. Large molecules are already so complex that they give experimental access to some aspects of the quantum-classical interface, i.e., to certain decoherence mechanisms. In 2011, the interference of molecules as heavy as 6910 u could be demonstrated in a Kapitza–Dirac–Talbot–Lau interferometer. In 2013, the interference of molecules beyond 10,000 u has been demonstrated.

Whether objects heavier than the Planck mass (about the weight of a large bacterium) have a de Broglie wavelength is theoretically unclear and experimentally unreachable; above the Planck mass a particle's Compton wavelength would be smaller than the Planck length and its own Schwarzschild radius, a scale at which current theories of physics may break down or need to be replaced by more general ones.

Recently Couder, Fort, *et al.* showed that we can use macroscopic oil droplets on a vibrating surface as a model of wave–particle duality—localized droplet creates periodical waves around and interaction with them leads to quantum-like phenomena: interference in double-slit experiment, unpredictable tunneling (depending in complicated way on practically hidden state of field), orbit quantization (that particle has to 'find a resonance' with field perturbations it creates—after one orbit, its internal phase has to return to the initial state) and Zeeman effect.

Treatment in Modern Quantum Mechanics

Wave–particle duality is deeply embedded into the foundations of quantum mechanics. In the formalism of the theory, all the information about a particle is encoded in its *wave function*, a complex-valued function roughly analogous to the amplitude of a wave at each point in space. This function evolves according to a differential equation (generically called the Schrödinger equation). For particles with mass this equation has solutions that follow the form of the wave equation. Propagation of such waves leads to wave-like phenomena such as interference and diffraction. Particles without mass, like photons, have no solutions of the Schrödinger equation so have another wave.

The particle-like behavior is most evident due to phenomena associated with measurement in quantum mechanics. Upon measuring the location of the particle, the particle will be forced into a more localized state as given by the uncertainty principle. When viewed through this formalism, the measurement of the wave function will randomly "collapse", or rather "decohere", to a sharply peaked function at some location. For particles with mass the likelihood of detecting the particle at any particular location is equal to the squared amplitude of the wave function there. The measurement will return a well-defined position, (subject to uncertainty), a property traditionally associated with particles. It is important to note that a measurement is only a particular type of interaction where some data is recorded and the measured quantity is forced into a particular eigenstate. The act of measurement is therefore not fundamentally different from any other interaction.

Following the development of quantum field theory the ambiguity disappeared. The field permits solutions that follow the wave equation, which are referred to as the wave functions. The term particle is used to label the irreducible representations of the Lorentz group that are permitted by the field. An interaction as in a Feynman diagram is accepted as a calculationally convenient approximation where the outgoing legs are known to be simplifications of the propagation and the internal lines are for some order in an expansion of the field interaction. Since the field is non-local and quantized, the phenomena which previously were thought of as paradoxes are explained. Within the limits of the wave-particle duality the quantum field theory gives the same results.

Visualization

There are two ways to visualize the wave-particle behaviour: by the "standard model", described below; and by the Broglie–Bohm model, where no duality is perceived.

Below is an illustration of wave–particle duality as it relates to De Broglie's hypothesis and Heisenberg's uncertainty principle (above), in terms of the position and momentum space wavefunctions for one spinless particle with mass in one dimension. These wavefunctions are Fourier transforms of each other.

The more localized the position-space wavefunction, the more likely the particle is to be found with the position coordinates in that region, and correspondingly the momentum-space wavefunction is less localized so the possible momentum components the particle could have are more widespread.

Conversely the more localized the momentum-space wavefunction, the more likely the particle is to be found with those values of momentum components in that region, and correspondingly the less localized the position-space wavefunction, so the position coordinates the particle could occupy are more widespread.

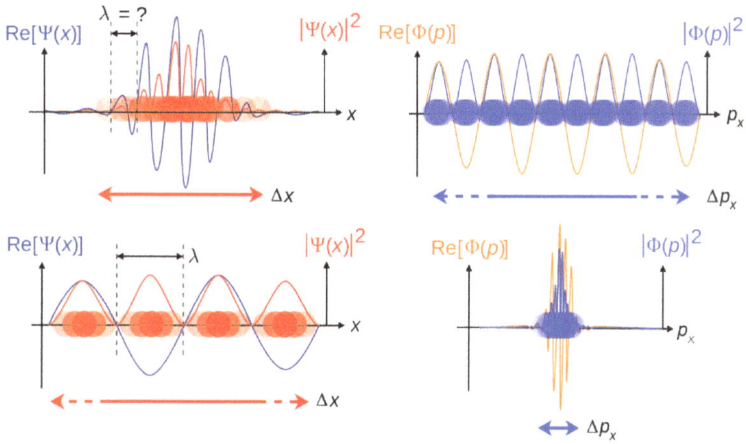

Position *x* and momentum *p* wavefunctions corresponding to quantum particles. The colour opacity (%) of the particles corresponds to the probability density of finding the particle with position *x* or momentum component *p*.

Top: If wavelength λ is unknown, so are momentum p, wave-vector k and energy E (de Broglie relations). As the particle is more localized in position space, Δx is smaller than for Δp_x.

Bottom: If λ is known, so are p, k, and E. As the particle is more localized in momentum space, Δp is smaller than for Δx.

Alternative views

Wave–particle duality is an ongoing conundrum in modern physics. Most physicists accept wave-particle duality as the best explanation for a broad range of observed phenomena; however, it is not without controversy. Alternative views are also presented here. These views are not generally accepted by mainstream physics, but serve as a basis for valuable discussion within the community.

Both-particle-and-wave View

The pilot wave model, originally developed by Louis de Broglie and further developed by David Bohm into the hidden variable theory proposes that there is no duality, but rather a system exhibits both particle properties and wave properties simultaneously, and particles are guided, in a deterministic fashion, by the pilot wave (or its "quantum potential") which will direct them to areas of constructive interference in preference to areas of destructive interference. This idea is held by a significant minority within the physics community.

At least one physicist considers the "wave-duality" as not being an incomprehensible mystery. L.E. Ballentine, *Quantum Mechanics, A Modern Development*, explains:

When first discovered, particle diffraction was a source of great puzzlement. Are "particles" really "waves?" In the early experiments, the diffraction patterns were detected holistically by means of a photographic plate, which could not detect individual particles. As a result, the notion grew that particle and wave properties were mutually incompatible, or complementary, in the sense that different measurement apparatuses would be required to observe them. That idea, however, was only an unfortunate generalization from a technological limitation. Today it is possible to detect the arrival of individual electrons, and to see the diffraction pattern emerge as a statistical pattern made up of many small spots (Tonomura et al., 1989). Evidently, quantum particles are indeed particles, but whose behaviour is very different from classical physics would have us to expect.

The Afshar experiment (2007) may suggest that it is possible to simultaneously observe both wave and particle properties of photons. This claim is, however, disputed by other scientists.

Wave-only View

At least one scientist proposes that the duality can be replaced by a "wave-only" view. In his book *Collective Electrodynamics: Quantum Foundations of Electromagnetism* (2000), Carver Mead purports to analyze the behavior of electrons and photons purely in terms of electron wave functions, and attributes the apparent particle-like behavior to quantization effects and eigenstates. According to reviewer David Haddon:

Mead has cut the Gordian knot of quantum complementarity. He claims that atoms, with their neutrons, protons, and electrons, are not particles at all but pure waves of matter. Mead cites as the gross evidence of the exclusively wave nature of both light and matter the discovery between 1933 and 1996 of ten examples of pure wave phenomena, including the ubiquitous laser of CD players, the self-propagating electrical currents of superconductors, and the Bose–Einstein condensate of atoms.

Albert Einstein, who, in his search for a Unified Field Theory, did not accept wave-particle duality, wrote:

This double nature of radiation (and of material corpuscles)...has been interpreted by quantum-mechanics in an ingenious and amazingly successful fashion. This interpretation...appears to me as only a temporary way out...

The many-worlds interpretation (MWI) is sometimes presented as a waves-only theory, including by its originator, Hugh Everett who referred to MWI as "the wave interpretation".

The *Three Wave Hypothesis* of R. Horodecki relates the particle to wave. The hypothesis implies that a massive particle is an intrinsically spatially as well as temporally extended wave phenomenon by a nonlinear law.

Particle-only View

Still in the days of the old quantum theory, a pre-quantum-mechanical version of wave–particle duality was pioneered by William Duane, and developed by others including Alfred Landé. Duane explained diffraction of x-rays by a crystal in terms solely of their particle aspect. The deflection of the trajectory of each diffracted photon was explained as due to quantized momentum transfer from the spatially regular structure of the diffracting crystal.

Neither-wave-nor-particle View

It has been argued that there are never exact particles or waves, but only some compromise or intermediate between them. For this reason, in 1928 Arthur Eddington coined the name "*wavicle*" to describe the objects although it is not regularly used today. One consideration is that zero-dimensional mathematical points cannot be observed. Another is that the formal representation of such points, the Dirac delta function is unphysical, because it cannot be normalized. Parallel arguments apply to pure wave states. Roger Penrose states:

"Such 'position states' are idealized wavefunctions in the opposite sense from the momentum states. Whereas the momentum states are infinitely spread out, the position states are infinitely concentrated. Neither is normalizable [...]."

Relational Approach to Wave–particle Duality

Relational quantum mechanics is developed which regards the detection event as establishing a relationship between the quantized field and the detector. The inherent ambiguity associated with applying Heisenberg's uncertainty principle and thus wave–particle duality is subsequently avoided.

Applications

Although it is difficult to draw a line separating wave–particle duality from the rest of quantum mechanics, it is nevertheless possible to list some applications of this basic idea.

- Wave–particle duality is exploited in electron microscopy, where the small wavelengths associated with the electron can be used to view objects much smaller than what is visible using visible light.
- Similarly, neutron diffraction uses neutrons with a wavelength of about 0.1 nm, the typical spacing of atoms in a solid, to determine the structure of solids.
- Photos are now able to show this dual nature, perhaps this will lead to new ways of examining and recording this behaviour.

Electromagnetic Radiation – Matter Interactions

- We have learnt that the radiation or light is following the Wave-Particle dual nature
- It can be treated as electromagnetic wave or particle nature like photon

- Similarly, the electrons can also be treated as particle and also as wave corresponding to its momentum for describing the stable structure of atom.
- When both of them interact with each other, we have to understand the mechanism to follow for describing the experimental observations such as Compton effect, absorption and emission of light by atoms.
- The different kinds of treatment to understand the light-matter interactions are described. We will start with the classical phenomenon and then proceed to understand the quantum mechanical descriptions.
- In classical theory as shown in figure (a), we model the atom as a heavy nucleus with electron attached to it with spring →the binding force between them.
- The resonant frequency for this system is ω_0
- If we treat the light as wave then the electric field can be represented as

$\vec{E} = \vec{E}_0 \sin \omega t$

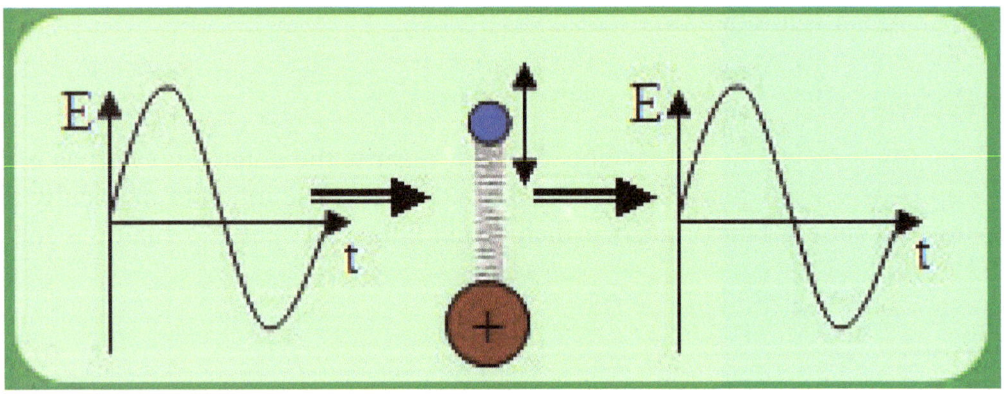

Figure (a)

- The oscillating electric field will force the electron to oscillate. The displacement of electron with respect to nucleus will produce an oscillating dipole.
- It is known that an oscillating dipole emit electromagnetic radiation with the same frequency of vibration. This emission of light is known as scattering of light by matter.
- However, if the incident radiation frequency matches with the resonant frequency of the system (ω_0) then the resonance occurs. Energy transfer takes place.

The oscillating dipole $p(t) = -ex(t)$

Where *e* is the charge of the electron and *x(t)* is the time dependent displacement of electron with respect to nucleus.

When this dipole oscillates, it emits radiation. As time passes by, due to the radiation loss the emission dies or decays. This phenomenon can be modeled as a damping oscillator whose solution will be x(t).

- The differential equation of motion of the damping oscillator with damping constant y

$\ddot{x}(t)+y\dot{x}(t)+\omega_0^2 x(t)=0$

Where $\omega_0^2 = {k}/{m}$, here k is the force constant of the spring.

With the initial values i.e. at t = o

the displacement $x(0)=x_0$ and the velocity $\dot{x}(0)=0$, the solution is

$$x(t)=x_0 e^{-({y}/{2\omega})t}[\text{Cos}\,\omega t+({y}/{2\omega})\sin\omega t]$$

The frequency $\omega=(\omega_0^2-\frac{\gamma^2}{4})^{1/2}$ of the damped oscillation is slightly lower than the frequency ω_0 of the undamped case.

Taking $y << \omega_0$, we get $x(t)=x_0 e^{-({y}/{2\omega})t}\cos\omega_0 t$.

Line Profile of the Emitted Radiation

The damped oscillation $x(t)$ can be described as a superposition of monochromatic oscillations with slightly different frequencies ω and amplitude $A(\omega)$

$$x(t)=\frac{1}{\sqrt{2\pi}}\int_{-\omega}^{+\omega}A(\omega)e^{i\omega t}d\omega$$

The amplitude $A(\omega)$ can be calculated by $A(\omega)=\frac{1}{\sqrt{2\pi}}\int_{-\omega}^{+\omega}x(t)e^{-i\omega t}dt$

The intensity

$$l(\omega-\omega_0)=A(\omega)A^*(\omega)=l_0\frac{1}{(\omega-\omega_0)^2+(y/2)^2}$$

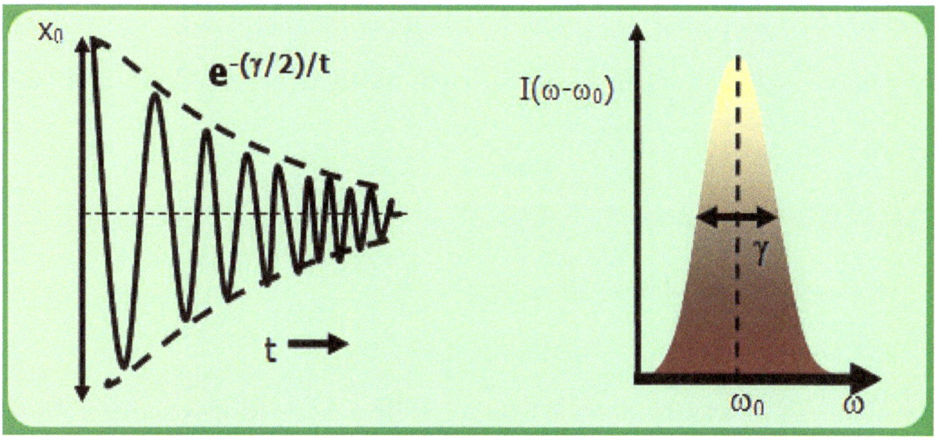

Figure (b)

Momentum of a Photon

Momentum vector magnitude $= |p| = \frac{E}{c}$

E is the energy.

$|p| = \frac{E}{c} = \frac{hv}{c} = h\lambda$ parallel to the direction of propagation.

From relativity, we know that

$$E = mc^2$$

$$p = mV$$

$$m = \frac{m_0}{\sqrt{1 - V^2/c^2}}$$

$$p = mV = \frac{E}{c^2} V$$

Minkowski's four dimensional space & time (E & three comp. of p)

If the system is confined to a single particle so, $E^2 - p^2c^2$ is invariant under changes of ref. frame.

$$E^2 - p^2c^2$$

$$= E^2 - \frac{E^2}{c^4} V^2 c^2 = E^2 \left(1 - \frac{V^2}{c^2}\right)$$

$$= m^2 c^4 \left(1 - V^2/c^2\right)$$

$$= \frac{m_0^2 c^4}{\left(1 - V^2/c^2\right)} \left(1 - V^2/c^2\right) = m_0^2 c^4$$

Photon is a quantity of energy. If we consider photon as a particle, then the energy $E = mc^2$, with this infinite energy E.

So the only way of reconciling a speed $V = c$ with a finite energy is to assume rest mass $m_0 = 0$.

So, $E^2 - p^2c^2 = 0$; $p = E/c$

Rest energy $= m_0 c^2 = 0$ for photon

So, $K.E. = E - m_0 c^2 = E$ so for photon all energy is K.E.

Elastic Collisions of Photons – Compton Effect

The X-ray of wavelength λ_0 was incident on a target. With a crystal diffractometer, the wavelengths of the scattered X-ray were measured by changing the angle q as shown in figure (c).

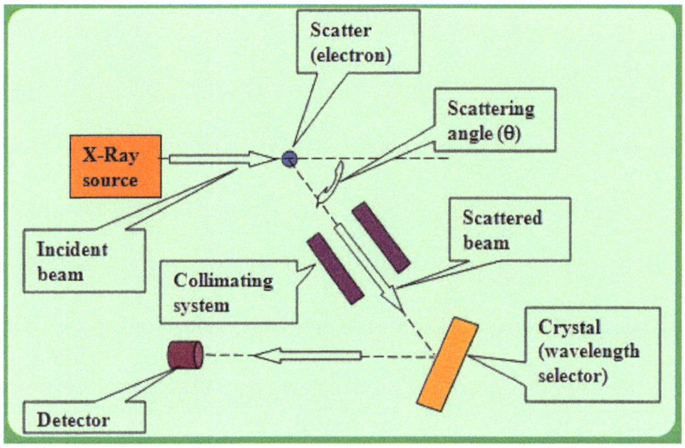

Figure (c)

Scattered radiation is composed by two lines:

i. a component at the incident wavelength λ_0, called the Thomson component. Scattered radiation has the same frequency as the incident radiation.

ii. a component of different wavelength $\lambda = \lambda_0 + \partial\lambda$.

Conclusion from experiment:

i. $\partial\lambda = \lambda - \lambda_0$ is always positive.

ii. $\delta\lambda$ is an increasing function of θ.

iii. $\delta\lambda$ is independent of and the composition of the material used for scattering.

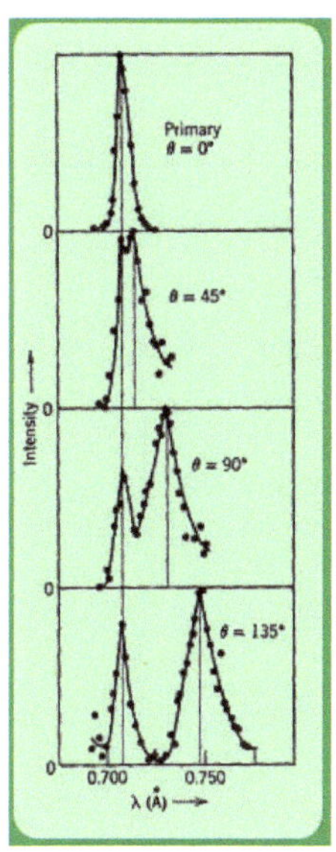

Elastic scattering between photons and free electrons (weakly bound to atoms can be considered as free electrons) i.e. binding energy $<< h\nu_0$.

Elastic Scattering $\rightarrow$ K.E. is conserved

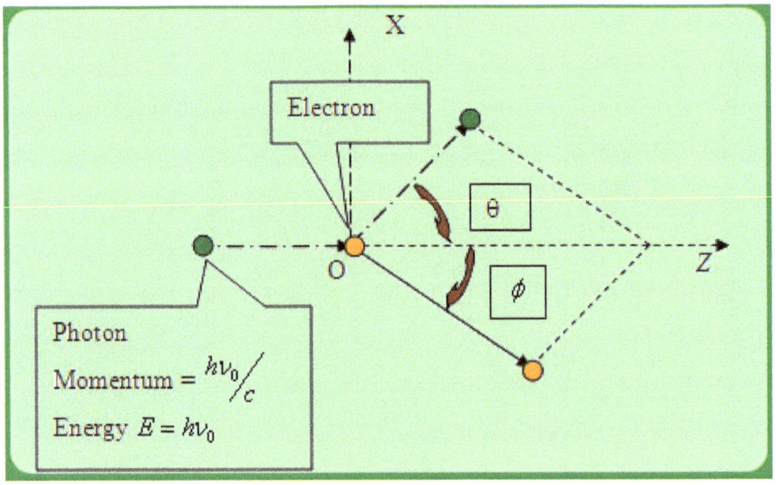

Figure (d)

We have taken into account the fact that p & E are related by relativistic invariance $E^2 - p^2c^2 = m_0^2c^4$.

	Before Collision	After Collision
Photon	Energy $hv_0 = h\, c/\lambda_0$	$hv = h\, c/\lambda$
	Momentum $h\, v_0/c = h/\lambda_0$ along Oz	$h\, v/c = h/\lambda$
Electron	Energy $= m_0c^2$	$\omega = mc^2$ $= \sqrt{(p^2c^2 + m_0^2c^4)}$
	Momentum = Zero	p making an angle ϕ with OZ

Conservation of Energy and Momentum

Energy $hv_0 + m_0c^2 + hv + \sqrt{(p^2c^2 + m_0^2c^2)}$

Momentum

$$alongOZ \quad \frac{hv_0}{c} = \frac{hv}{c}\cos\theta + p\cos\phi$$

$$alongOX \quad 0 = \frac{hv}{c}\sin\theta + p\sin\phi$$

$$\cos\phi = \frac{hv}{c} - \frac{hv}{c}\cos\theta$$

$$= -(v \quad - v\cos \quad)$$

$$p\sin\phi = -- v\sin\theta$$

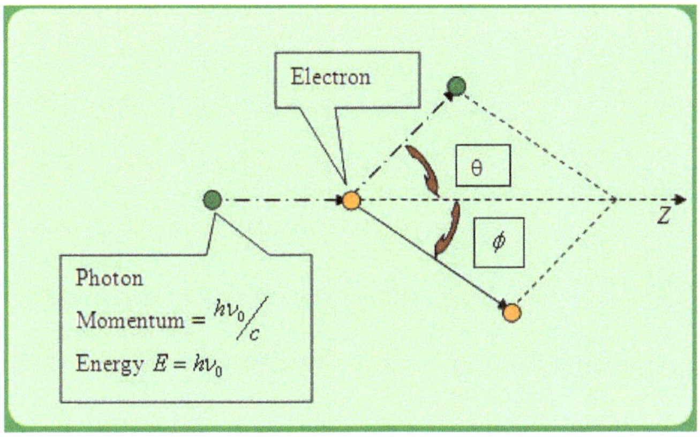

Figure (e)

$$p^2 = \frac{h^2}{c^2}(v_0 - v\cos\theta)^2 + \frac{h^2}{c^2}v^2\cos^2\theta$$

$$= \frac{h^2}{c^2}(v_0^2 + v^2\cos^2\theta - 2vv_0\cos\theta + v^2\sin^2\theta$$

$$= \frac{h^2}{c^2}(v_0^2 + v^2 - 2vv_0\cos\theta)$$

$$\Rightarrow p^2c^2 = h^2(v_0^2 + v^2 - 2vv_0\cos\theta)$$

From 1st equation $p^2c^2 = [h(v_0 - v) + m_0c^2]^2 - m_0^2c^2$

Equating the equations above

$$h^2(v_0^2 + v^2 - 2vv_0\cos\theta) = \left[h(v_0 - v) + m_0c^2\right]^2 - m_0^2c^4$$

$$= h^2(v_0 - v)^2 + \cancel{m_0^2c^4} + 2h(v_0 - v)m_0c^2 - \cancel{m_0^2c^4}$$

$$\Rightarrow \cancel{h^2v_0^2} + \cancel{h^2v^2} - 2h^2vv_0\cos\theta = \cancel{h^2v_0^2} + \cancel{h^2v^2} - 2h^2vv_0 2h(v_0 - v)m_0c^2$$

$$\Rightarrow 2h(v_0 - v)m_0c^2 = 2h^2(1 - \cos\theta)vv_0$$

$$\Rightarrow v_0 - v = \frac{\cancel{2h^2}}{\cancel{2h}m_0c^2}(1 - \cos\theta)vv_0$$

$$\Rightarrow \frac{v_0}{vv_0} - \frac{v}{vv_0} = \frac{h}{m_0c^2}(1 - \cos\theta) \qquad where\ m_0 = 9.1\times10^{-31}kg$$

$$\Rightarrow \frac{1}{v} - \frac{1}{v_0} = \frac{h}{m_0c^2}(1 - \cos\theta)$$

$$\Rightarrow \lambda - \lambda_0 = \frac{h}{m_0c}(1 - \cos\theta)$$

$$\Rightarrow \delta\lambda = \frac{h}{m_0c}(1 - \cos\theta) \quad \begin{cases} \theta = 0; & \delta\lambda = 0 \\ \theta = \pi; & \delta\lambda = \dfrac{2h}{m_0c} \\ \theta = 90^\circ; & \delta\lambda = \dfrac{h}{m_0c} = \wedge = Compton\ wavelength \end{cases}$$

So, we get

$$\lambda - \lambda_0 = \partial\lambda = \frac{h}{m_0 c}(1-\cos\theta)$$

i. $\partial\lambda$ is positive.

ii. independent of λ_0.

iii. depends on θ only.

Hard X-ray $\lambda = 1pm$

$hv = 1MeV$

Energy of Compton Wavelength,

$$\frac{hc}{\Lambda} = \frac{hc}{\frac{h}{m_0 c}} = m_0 c^2 \quad \Rightarrow \textit{rest mass of the electron}$$

$hv_0 << m_0 c^2$ $\lambda_0 >> \Lambda$ $\partial\lambda << \lambda_0$ photons gives up very little energy.

$hv_0 > m_0 c^2$ $\lambda_0 < \Lambda$ $\partial\lambda > \lambda_0$ photons gives up most of its energy.

Absorption of Photons (Inelastic Collision)

Let us define hv_{12} is the energy difference between the two states of an atom with energy E_2 and E_1 ($E_2 > E_1$ ground state) $E_2 - E_1 = hv_{12}$

If there is a collision between an atom and a photon as shown in figure (f) and after the collision they form one particle, then the initial momentum of the incident must be preserved by the final single particle (momentum conservation). Or we can say that after collision the atom must possess some kinetic energy that it did not have previously.

This kinetic energy can have been taken from the energy of the incident photon.

This is possible only when $hv > hv_{12}$. We have to determine the relationship between v and v_{12}.

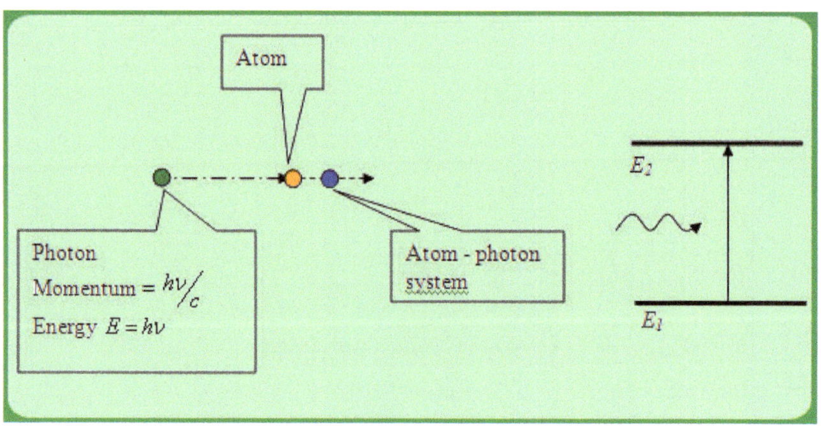

Figure (f)

Considering the properties of the relativistic particles:

		Before Collision	After Collision
Photon	Energy	hv	
	Momentum	hv/c	
Atom	Energy	$E_1 = m_1 e^2$ $m_1 \rightarrow rest\ mass$	$W = \sqrt{(p^2c^2 + m_2^2e^2}$
	Momentum	Zero	p $where, m_2 = \frac{E_2}{c^2} = \frac{E_1 + hv_{12}}{c^2}$

From the conservation of energy

$$hv + E_1\sqrt{(p^2c^2 + m_2^2e^4)}$$

$$Momentum\ p = hv/c\ same\ as\ photon$$

$$E_1 + hv_{12} = m_2c^2$$

From Equations above,

$$(E_1 + hv)^2 = p^2c^2 + m_2^2c^2$$

$$\Rightarrow E_1^2 + 2E_1hv + h^2v^2 = p^2c^2 + (E_1 + hv_{12})^2$$

$$\frac{h^2v^2}{c^2}c^2 + E_1^2 + 2E_1hv_{12}h^2v_{12}^2$$

$$\Rightarrow v = \frac{2hv_{12}E_1}{2E_1h}\left[1 + \frac{h^2v_{12}^2}{2E_1hv_{12}}\right] = v_{12}\left[1 + \frac{hv_{12}}{2E_1}\right]$$

$$= v_{12}\left[1 + \frac{hv_{12}}{2m_1c^2}\right]$$

So the incident photon frequency must be greater than the theoretical frequency of the spectral lines

$$\left(\frac{E_2 - E_1}{h} = v_{12}\right)$$

$$\left.\begin{array}{ll} hv_{12} & 1\ to\ 10eV \\ m_1c^2 & 10\ to\ 100GeV \end{array}\right\} \quad 10^{-10}$$

Comment

i. In the optical transitions $h\nu_{12} << m_1c^2$ typically, $\frac{h\nu_{12}}{m_1c^2} \sim 10^{-10}$ which means the shift will be very small compared to other broadening (10^{-6}) cannot be observed. Resonant condition is applicable i.e. $E_2 - E_1 = h\nu$.

ii. When $h\nu_{12}$ is not very small compared to m_1c^2, the shift is appreciable.

Emission of Photons

		Before Collision	After Collision
Photon	Energy	~	$h\nu$
	Momentum	~	$h\nu/c$ along Oz
Atom	Energy	E_2	$W = \sqrt{(p^2c^2 + m_1^2c^4)}$
	Momentum	0 $m_2\, E_2/c^2$	p along Oz $m_1 = E_1/c^2 = \frac{E_2 + h\nu_{12}}{c^2}$

$$p + \frac{h\nu}{c} = 0 \;\Rightarrow p^2c^2 = h^2\nu^2$$

$$E_2 = W + h\nu$$

$$\Rightarrow E_2 - h\nu = W$$

$$\Rightarrow E_2^2 + h^2\nu^2 - 2h\nu E_2 = p^2c^2 + m_1^2c^4$$

$$h^2\nu^2 + m_1^2c^4$$

$$\Rightarrow E_2^2 - 2h\nu E_2 = m_1^2c^4 = (E_2 - h\nu_{12})^2$$

$$\Rightarrow E_2^2 - 2h\nu E_2 = E_2^2 + h^2\nu_{12}^2 - 2E_2h\nu_{12}$$

$$\Rightarrow 2h\nu E_2 = 2E_2h\nu_{12} - h^2\nu_{12}^2$$

$$\Rightarrow \nu = \frac{2E_2h\nu_{12}}{2hE_2} - \frac{h^2\nu_{12}^2}{2E_2h} = \nu_{12} - \frac{h\nu_{12}^2}{2E_2}$$

$$\Rightarrow \nu = \nu_{12}\left[1 - \frac{h\nu_{12}}{2m_2c^2}\right]$$

Application to γ–rays

$$y - rays \rightarrow Very\ High\ Energy$$

$$10^{19}\ to\ 10^{22}\ Hz$$

Nuclear transitions also very high frequency

$$h\nu_{12} < m_1c^2$$

Emission of $y-ray$ photon is interpreted as the radiative transition between two energy states of the nucleus (analogous to optical transition).

$Frequency\ Displacement > Line\ width$

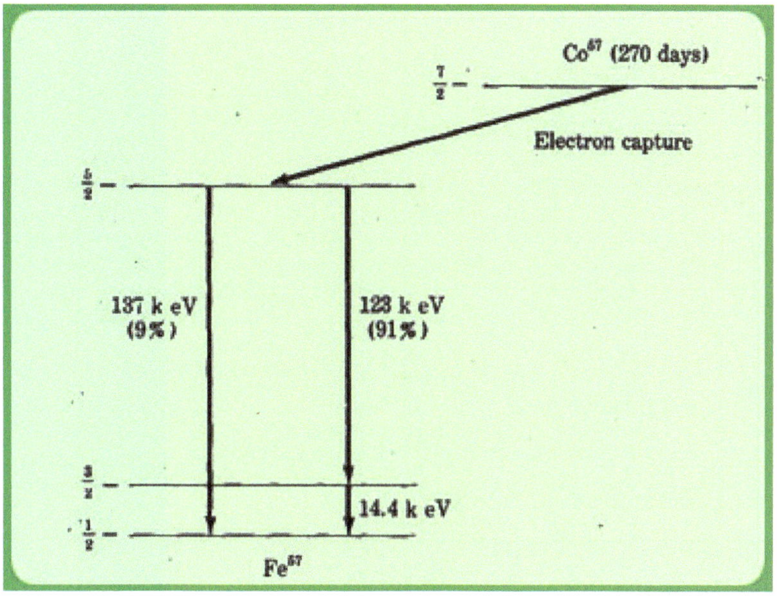

Figure (g)

Resonance absorption and emission is not possible as in the case of optical transition. By increasing temperature it is possible. So to get the information about the nucleus from this is difficult.

What Mössbaur discovered that by cooling the source and the absorber the intensity absorption increased.

It means that, below a certain temperature emitted or absorbing nucleus is embedded in a crystalline lattice and it is the white crystal which recoils.

$Mass\ of\ crystal >> Mass\ of\ Nucleus \sim 6\times10^{23}$

The velocity of recoil is negligible.

No recoil energy loss $\rightarrow$ extremely sharp lines (natural width) could be obtained.

$Debye-Waller\ factor$

$$f = \exp\left[-\frac{E_0}{K\Theta}\left(\frac{3}{2}+\frac{\pi^2T^2}{\Theta}\right)\right]$$

$where, T = Abs.temp.$

$K = Boltzmann's\ const.$

$\Theta = Degree\ Temp.\ of\ the\ solid$

There are three main hyperfine interactions that can be observed by Mössbauer spectroscopy. They are (i) Isomer Shift, (ii) Quadrupole Splitting and (iii) Nuclear Zeeman Splitting.

For example :

The isomer shift of $Fe^{57}(3/2 \rightarrow 1/2 = 14.4KeV)$ in Ferricinium bromide is $+2\times10^{-8}eV$.

$$h(v-v_0) = h\partial v = 2\times10^{-8}eV$$

$$hv_0 = 14.4\times10^3 eV$$

$$\frac{v}{v_0} = \frac{c}{c-v}$$

$$\Rightarrow \frac{\delta v}{v_0} = \frac{v}{c-v} \approx \frac{v}{c}$$

$$\Rightarrow v = \frac{\partial v}{v_0}\times c = \frac{2\times10^{-8}}{14.4\times10^3}\times3\times10^{10}$$

$$= \frac{6\times10^2}{14.4\times10^3} = 0.04\,cm/s$$

Mössbaur Spectrum of Ferricinium bromide

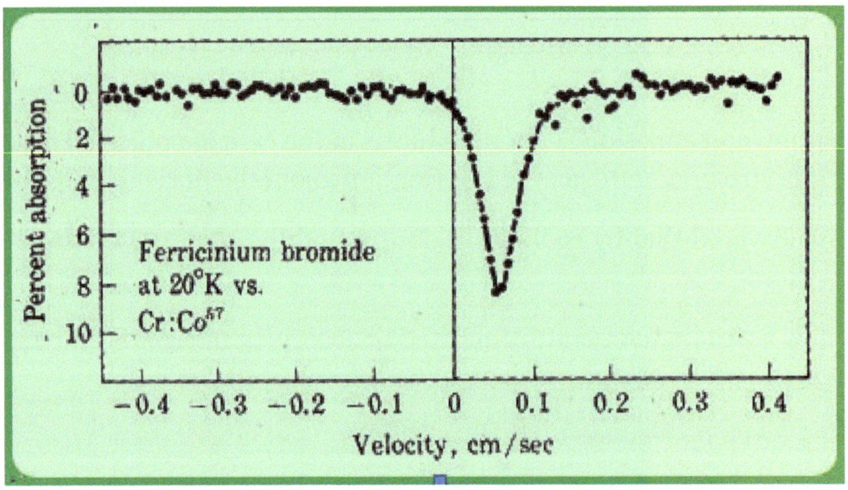

Radiative Transitions and Spectral Broadening

- The spectral lines emitted by atomic vapors at moderate temperature and pressure show the wavelength spread around the central frequency. This is known as width of the spectral lines.
- First we will understand the origin of radiative transitions.
- We will establish the relations between Einstein coefficients B_{12}, B_{21} and A_{21}
- We will discuss the different mechanism for the broadening of the spectral lines emitted by the atomic source. We will find out the reasons for these broadening.

Radiative Transitions and Spectral Broadening

- We now know that, Bohr's theory described that the emission of radiation from atoms was from the higher energy state to lower energy state of atom.
- This is known as "quantum jump"
- Bohr tried to describe it through the correspondence principle

According to Bohr's theory,

$$E_{n+1} = -\frac{2\pi^2 me^4}{(n+1)^2 h^2}$$

And $E_n = -\frac{2\pi^2 me^4}{(n)^2 h^2}$

Thus $E_{n+1} - E_n = \frac{2\pi^2 me^4}{h^2}[\frac{1}{(n)^2} - \frac{1}{(n+1)^2}]$

$$\approx \frac{2\pi^2 me^4}{h^2}\frac{2}{n^3} \text{ for } n >> 1$$

The emitted frequency $E_{n+1} - E_n = h\nu_{n+1,n}$

However, Bohr's theory fails to describe the mechanism of transitions between the two stationary states. It also does not provide rational explanation for deriving the intensity and polarization of the emitted spectral lines.

Einstein Transition Probabilities

The first major step to understand the transitions between two stationary states and the corresponding radiation was taken by Einstein.

As shown in figure below, let us assume that E_1 and E_2 are the stationary energy levels of an atomic system. Let us consider that this system is exposed to a radiation $h\nu$.

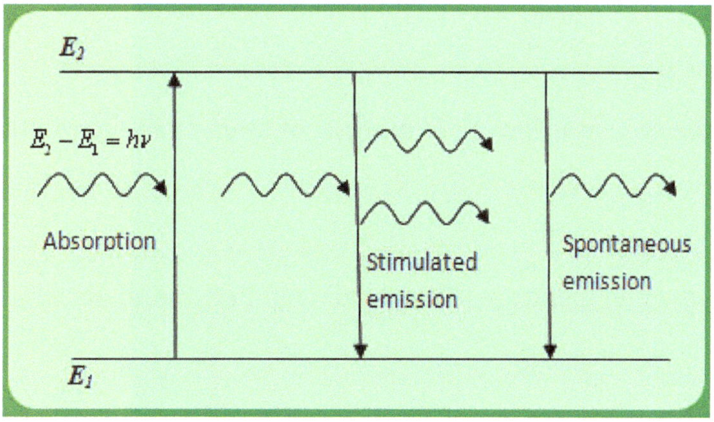

Process 1: If an atom absorbs a photon of energy $E_2 - E_1 = h\nu$, it is excited from the lower energy level *E1* to the higher energy level *E2*. The process is called induced absorption. The probability per second that the atom will absorb the photon, $\frac{dp_{12}}{dt}$

This is proportional to the number of photons of energy $h\nu$ per unit volume = the spectral density of the radiation field $\rho_\nu(\nu)$

So, $\frac{dp_{12}}{dt} = B_{12}\rho(\nu)$, B_{12} = Einstein coefficient of induced absorption.

Process 2: The radiation field can induce the atoms to make a transition from excited states E_2 to lower energy state E_1.

Since the system decreases the energy, the extra energy will be released as the emission of a photon of energy hv. This process is known as induced emission or stimulated emission.

The induced photon of hv will have the same nature as the photon caused this emission.

The probability $\frac{dp_{12}}{dt}$ that one atom emits one induced photon per second is

$$\frac{dp_{21}}{dt} B_{21}\rho(\nu)$$

B_{21}= Einstein coefficient of induced emission.

Process 3: An excited atom in the excited state E_2 can spontaneously jump into the lower energy states $E_1$by emitting a photon of energy $E_2 - E_1 = h\nu$

Note: Spontaneous radiation can be emitted in any arbitrary direction. The probability per second depends on the nature of the stationary energy state and independent of the external field.

So, $\frac{dp_{21}^{spont}}{dt} = A_{21}$

A_{21} = Einstein coefficient of spontaneous emission and also known as spontaneous transition probability.

So, $\frac{dp_{12}}{dt} = B_{12}\rho(\nu)$, B_{12} = Einstein coefficient of induced absorption.

Relation between B_{12}, B_{21} and A_{21}

Let us assume that the system is having *N* number of atoms and distributed in different energy levels *Ei* such that

$$\Sigma N_i = N$$

At thermal equilibrium of temperature (*T*), according to Boltzmann distribution,

$$N_i = N_0 g_i e^{-Ei/kT}$$

Where g_i is the degeneracy of level E_i.

In a stationary field,

The number of photons absorbed per unit volume per second = the emission state

$$B_{12}N_1\rho(v) = [B_{21}\rho(v) + A_{21}]N_2$$

Using

$$\frac{N_2}{N_1} = \left(\frac{g_2}{g_1}\right)e^{-(E_2-E_1)/kT} = \left(\frac{g_2}{g_1}\right)e^{-hv/kT}$$

We get,

$$\rho(v) = \frac{A_{21}/B_{21}}{\left(\frac{g_1}{g_2}\right)\left(\frac{B_{12}}{B_{21}}\right)e^{hv/kT-1}}$$

The radiation density given by plank's law,

$$\rho(v) = \frac{8\pi hv^3}{c^3}\frac{1}{e^{hv/kT}-1}$$

Equating we get,

$$B_{12} = \left(\frac{g_2}{g_1}\right)B_{21}$$

It implies that with the equal weights $g_2 = g_1$, the probability of induced emission is equal to that of induced absorption.

We also get

$$A_{21} = \frac{8\pi h^3}{c^3}B_{21}$$

This provides the number of modes per unit volume and unit frequency interval

$$n(v) = \frac{8\pi h^3}{c^3} \text{ and thus, } \frac{A_{21}}{n(v)} = B_{21}hv$$

This means that the ratio of the induced to the spontaneous emission rate in an arbitrary mode in equal to the number of photons in this mode.

Relation between Lifetime and Spontaneous Transition Probability

An excited atom in the excited state E_2 can spontaneously jump into the lower energy states E_1 by emitting a photon of energy $E_2 - E_1 = hv$

So,

$$\frac{dp_{21}^{spont}}{dt} = A_{21}$$

When there are several pathways then

$$A_i = \sum_k A_{ik}$$

The population of the excited state will decrease.

So

$$dN_i = -A_i N_i dt$$

We get

$$N_i(t) = N_i(0)e^{-Ait}$$

Where $N_i(0)$ is the population density at $t = o$

The population $N_i(0)$ will be $N_i(0)/e$ after time $\tau_i = 1/A_i$ as shown in the figure below. This τ_i represents the mean spontaneous lifetime of the level Ei.

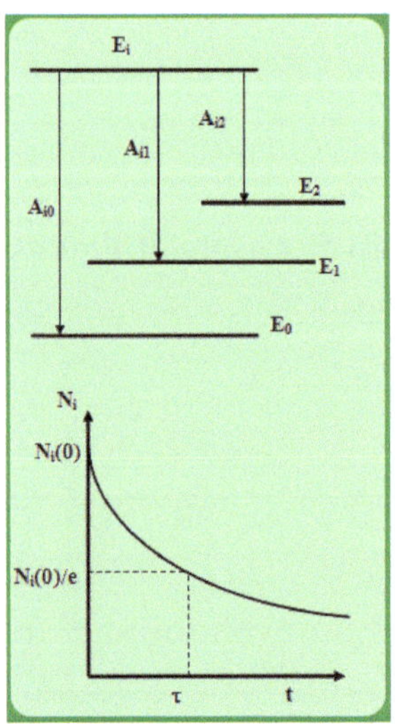

Semiclassical approach of radiative transitions:

In this approach, we consider the radiation as classical electromagnetic wave

$$\vec{E} = \vec{E}_0 \, \mathrm{Cos}(\omega t - kz)$$

On the other hand, we treat the atom quantum mechanically. That means, the atoms absorb or emit radiation when they jump between the two quantized states.

Since the dimension of the atom is much smaller (~0.5 nm) than the wavelength of light (~500 nm), we take

$$\vec{E} = \vec{E}_0 \cos(\omega t) = \frac{\vec{E}_0}{2}(e^{i\omega t} + e^{-i\omega t})$$

In the dipole approximation, the interaction energy $V = \vec{p}.\vec{E} = pE_0 \cos(\omega t)$ where the dipole $\vec{p} = -e\vec{r}$

The general solution $\Psi(r,t)$ of the time dependent Schrodinger Equation

$$H\Psi(r,t) = i\hbar \frac{\delta \Psi(r,t)}{\delta t}$$

Can be expressed as $\Psi(r,t) = \sum_m C_m(t)\phi_m(r)e^{-E-i/\hbar}$

Where $\phi_m(r)$ is the eigen function of the time independent Schrodinger equation

$$H\phi_m(r) = E_m\phi_m(r)$$

For two level system as shown in the figure below, $\Psi(r,t) = a(t)\phi_Q(r)e^{-E\sigma t/\hbar} + b(t)\phi_\delta(r)e^{-iE\sigma t/\hbar}$

Where *a(t)* and *b(t)* are the time dependent probability amplitude of the states *a* and *b* and

$$|a(t)|^2 + |b(t)|^2 = 1$$

Substituting in the Schrodinger equation, we get

$$i\hbar \frac{da(t)}{dt}\phi_Q e^{-iE\sigma t/\hbar} + i\hbar \frac{db(t)}{dt}\phi_\delta e^{-E\delta t/\hbar} = a(t)V\phi_Q e^{-iE\delta t/\hbar} + b(t)V\phi_\delta e^{-iE\delta t/\hbar}$$

Integrating over the space we get

$$\frac{da(t)}{dt} = -\frac{i}{\hbar}[a(t)V_{aa} + b(t)e^{i(E_\sigma - E_b)t/\hbar}]$$

And

$$\frac{db(t)}{dt} = -\frac{i}{\hbar}[b(t)V_{\delta\delta} + a(t)e^{-i(E_\sigma - E_b)t/\hbar}]$$

Where $V_{ab} = \int \phi_\sigma^\phi V \phi_\delta d\tau = -eE\int \phi_\sigma^\phi r \phi_b d\tau$

The dipole matrix element $D_{ab} = -e\int \phi_\sigma^\phi r \phi_\delta d\tau$

Since *r* has odd parity, we get $D_{ab} = D_{ba} = 0$ and $D_{ab} = D_{ba} = -e\int \phi_\sigma^\phi r \phi_\delta d\tau$

This depends on the stationary state wavefunctions $\phi_a(r)$ and $\phi_\delta(r)$ and need to calculate for understanding the transitions between the two states.

Intensity of the transition $I \propto |D_{ab}|^2$

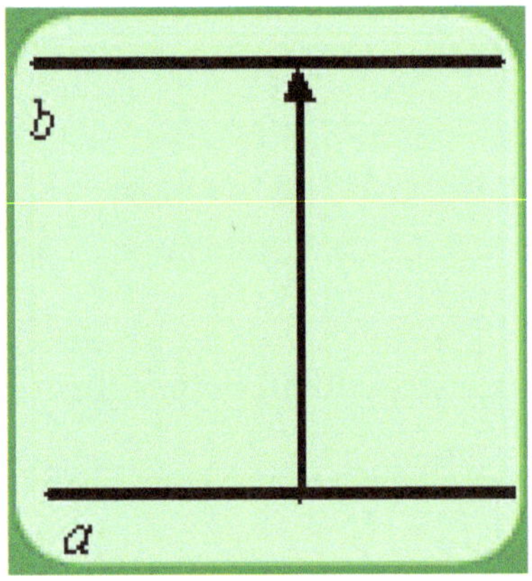

Widths and Profiles of Spectral Lines

As shown in the figure below, the spectral lines observed from atomic vapors are never strictly monochromatic.

Even with very high sensitive instrument, the observed lines show a wavelength spread around the line center.

This is known as width of the spectral lines.

Width of the spectral lines

Central frequency $\nu_0 = (E_a - E_b)/\hbar$ corresponding to a transition

The line profile $I(\nu)$ around ν_0 is known as line profile

ν_1 and ν_2 where the intensity is the half of the maximum intensity ($(I_0/2)$

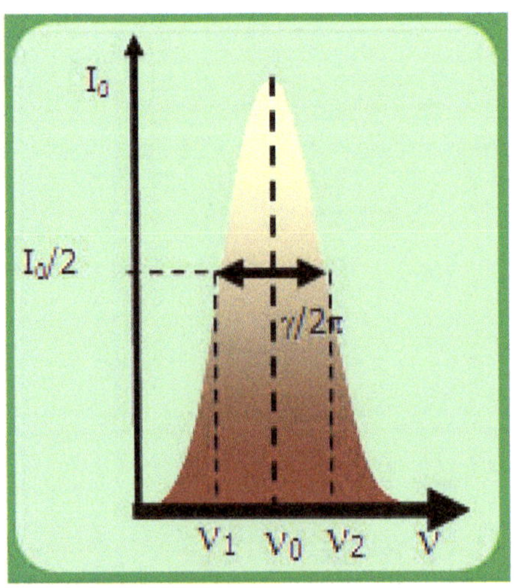

$\delta\nu = \nu_2 - \nu_1$ is known as the Full Width at Half Maximum (FWHM)

Since $\lambda = \dfrac{c}{\nu}$ then $\delta\lambda = -\left(\dfrac{c}{\nu^2}\right)\delta\nu$

The relative half width

$$\left|\frac{\delta v}{v}\right| = \left|\frac{\delta \lambda}{\lambda}\right|$$

Mainly there are two reasons for the broadening of the spectral lines

a. Width caused by the atomic source

b. Width arises due to the instrumental limitations

(a) There are basic three reasons for the broadening caused by the atomic source

1. Natural line broadening
2. Doppler line broadening
3. Pressure / Collisional line broadening

Depending on the effect on the individual atoms, there are two types of broadening.

If all the atoms are experiencing the same type of effect, then the broadening caused by this is known as Homogeneous broadening.

On the other hand, if individual atoms are experiencing different effect then this is known as inhomogeneous broadening.

1. Natural line broadening

An excited atom can emit the radiation spontaneously. This phenomenon can be treated classically.

The intensity profile

$$I(\omega - \omega_0) = A(\omega)A^*(\omega) = I_0 \frac{1}{(\omega - \omega_0)^2 + (y/2)^2}$$

Where Io is the maximum intensity and $\omega = 2\pi v$. This line profile is known as Lorentzian profile and can be written as

$$L(\omega - \omega_0) = \frac{1}{2\pi} \frac{y}{(\omega - \omega_0)^2 + (y/2)^2}$$

In this, full width half maximum (FWHM) = $\delta\omega = y$ or $\delta v = {}^{y}\!/_{2\pi}$

This also can be understood from the Heisenberg Uncertainty Principle

$\tau \Delta E = {}^{h}\!/_{2\pi}$ where τ is the lifetime and ΔE is the uncertainty in energy. So, when τ is small, ΔE is large.

This broadening is known as natural linewidth of the spectral line for the excited state whose lifetime is finite. The broadening of the spectral line can not be smaller than this ΔE even a high resolution instrument is used to measure it. This type of bradening is Homogeneous broadening.

$$Natural\ Width \propto \frac{1}{Life\ Time\ of\ Excited\ states}$$

At high pressure, due to collisions life of the excited state is decreased and the broadening occurs.

$$\tau = \frac{1}{n_0}\sum_{i=1}^{n_0} t_i \quad t_i \Rightarrow time\ for\ atom\ i\ in\ the\ excited\ state$$

↑

lifetime of the

excited state

If collision occurs it looses some excitation energy and for that if t_i decreases then the natural width increased.

For example: If the lifetime of a particular excited state is 10-9 sec then the energy broadening from Heisenberg uncertainty principle

$$\Delta E \Delta t = \hbar$$

$$\Rightarrow \Delta E = \frac{\hbar}{\Delta t} = \frac{1.05\times10^{-34}}{10^{-9}} J = 1\times10^{-25} J$$

2. Doppler line broadening

Atoms emitting electromagnetic radiation are not stationary (gas). For an observer in the laboratory frame of reference, the emission must be considered as coming from a source in motion, so necessary to take Dopper effect.

Speed V of atom is small compared to c, so it is possible to use classical expression for the Doppler effect.

Let θ be the angle between observation direction & velocity vector, the change of frequency $\delta\nu$ between the exact frequency ν_0 and the frequency ν seen by the observer,

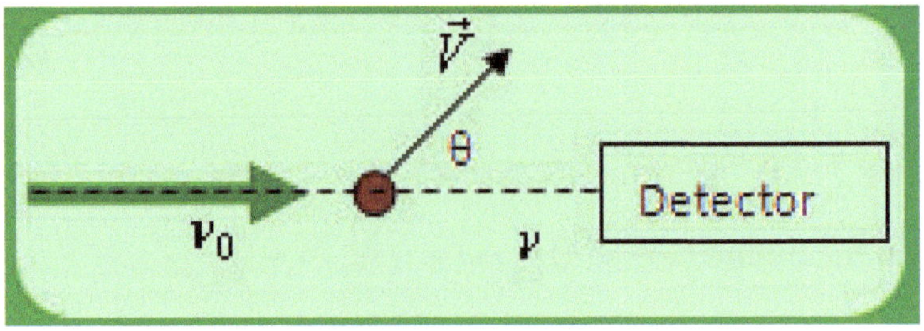

$$\frac{\delta\nu}{\nu_0} = \frac{\nu-\nu_0}{\nu_0} = \frac{V\cos\theta}{c} = \frac{V_x}{c}$$

V_x is the component in the direction of the observer. If we assume that the temperature of the gaseous source of light is uniform, the distribution of speeds of the atoms is a Maxwell distribution. So the number of atoms whose velocity V_x is between V_x and $V_x + dV_x$

$$dN = Nf(V_x)dV_x \qquad N \Rightarrow Total\ no.\ of\ Atoms$$

$f(V_x)$ is the probability density for the component v_x

$$f(V_x) = \sqrt{\left(\frac{M}{2\pi RT}\right)}\exp\left[-\frac{M}{2RT}V_x^2\right]$$

$M \Rightarrow$ molecular weight, $R \Rightarrow$ perfect gas constant

Now, $V_x = c\frac{v - v_0}{v_0}\ so, dV_x = c\frac{dv}{v_0}$

Let $p_v dv$ is the power emitted in the band frequencies between v and $v + dv$ is proportional to the number of atoms between velocity V_x and $V_x + dV_x$, if proportionality constant is K,

$$p_v dv = KNf(V_x)dV_x$$

$$= KNf\left(c\frac{v-v_0}{v_0}\right)c\frac{dv}{v_0}$$

$$= K'f\left(c\frac{v-v_0}{v_0}\right)dv$$

$$so, p_v = K'f\left(c\frac{v-v_0}{v_0}\right) = K''\exp\left[-\frac{M}{2RT}\frac{c^2(v-v_0)^2}{v_0^2}\right]$$

This profile as given in Equations above & as shown in the figure below is the Gaussian line profile.

$\Delta v_D = v_2 - v_1$, the width at half height of the curve.

$$\exp\left[-\frac{M}{2RT}\frac{c^2(v-v_0)^2}{v_0^2}\right] = \frac{1}{2}$$

$$\Rightarrow v - v_0 = \delta v = \frac{v_0}{c}\sqrt{\left(\frac{2RT}{M}In2\right)}$$

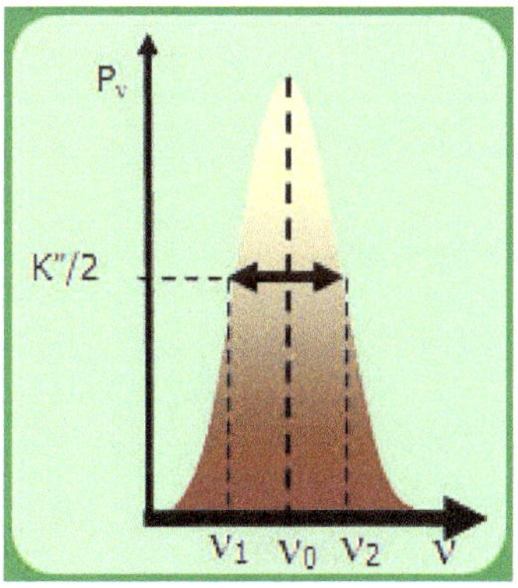

So the width $\Delta\nu_D = \frac{2\nu_0}{c}\sqrt{\left(\frac{2RT}{M} In2\right)}$

So, Doppler Width

$$\Delta\nu_D \propto \sqrt{T}$$
$$\propto \nu_0$$
$$\propto \frac{1}{\sqrt{M}}$$

$$\lambda = 589.3nm \rightarrow \Delta\left(\frac{1}{\lambda}\right) = 0.057cm^{-1}$$

Na gas at 500 *K*

$$\frac{v}{c} \approx 10^{-6} \approx \frac{\partial v}{v}$$
$$\frac{\delta v}{v} \approx 10^{-10}$$

Natural Line Width

So the Doppler width is much greater than the Natural line width. Doppler broadening is inhomogeneous broadening.

3. Pressure / Collisional line broadening

For a gas at a given pressure radiating atoms interact with the neighboring atoms via collision and this affect the emission line width strongly.

For example, an atom A with energy levels *E1* and *E2* approaches to another atom B. Because of the interaction, the energy levels get perturbed and the shift of these energy levels occur. Due to this the line profile gets broadened. This broadening depends on the collision diameter.

FWHM = $\frac{1}{\pi\tau_0}$ where τ_0 is the mean flight time between two successive collision.

(b) Width Arises Due to the Instrumental Limitations

The basic set up for the emission experiment is

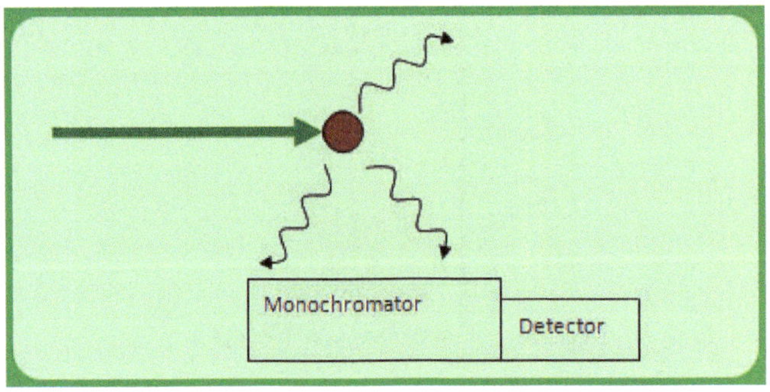

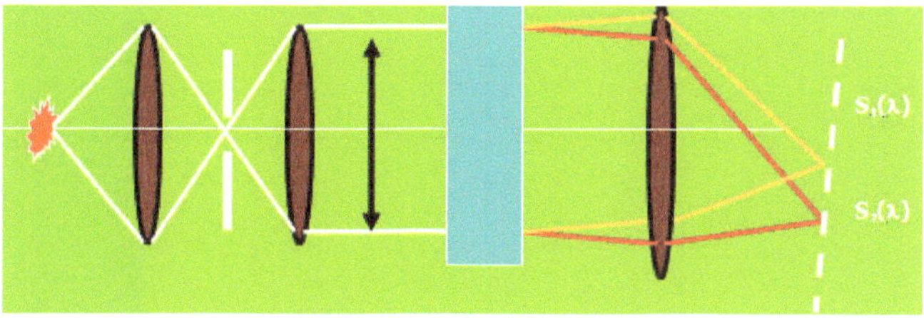

Source f_1 slit aperture Dispersing element f_2 Screen

Linear Dispersion = $\dfrac{dx}{d\lambda} \propto f(\dfrac{dn}{d\lambda}, f_2)$

Slit width = dx1

Image width = dx1(f2/f1)

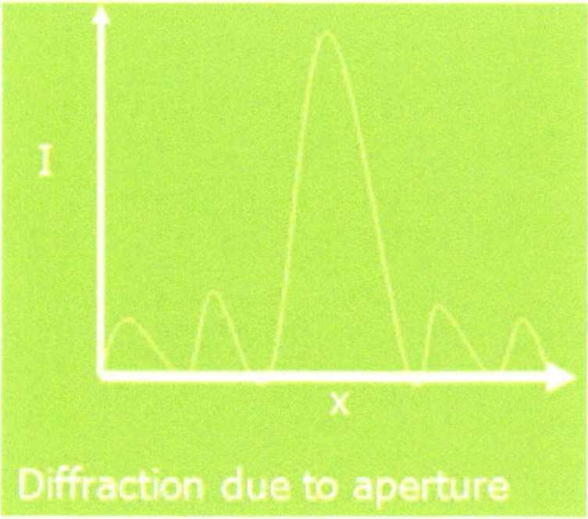

Spectral Resolving Power

$$I(\lambda-\lambda_0) = I(\lambda_0)[\sin((\lambda-\lambda_0)/2)(\lambda-\lambda_0)/2]^2$$
$$I(\lambda) = I_1(\lambda-\lambda_1) + I_2(\lambda-\lambda_2)$$

Rayleigh criterion0.8 of I_{max}

$$\Delta x_2 = f_2(\lambda / a)$$

Where a is the aperture

And the resolving power = $\dfrac{\lambda}{\delta\lambda}$

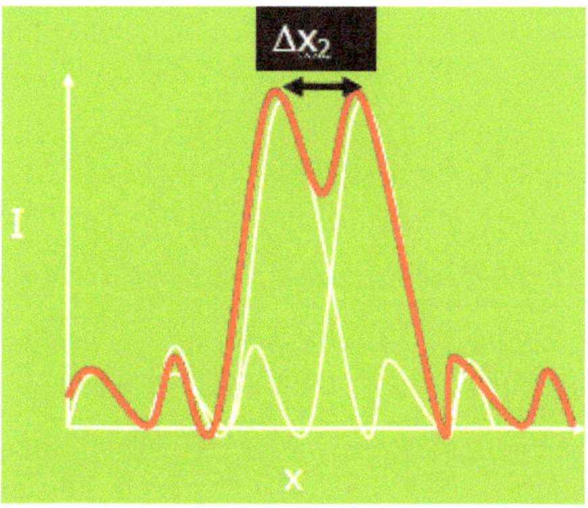

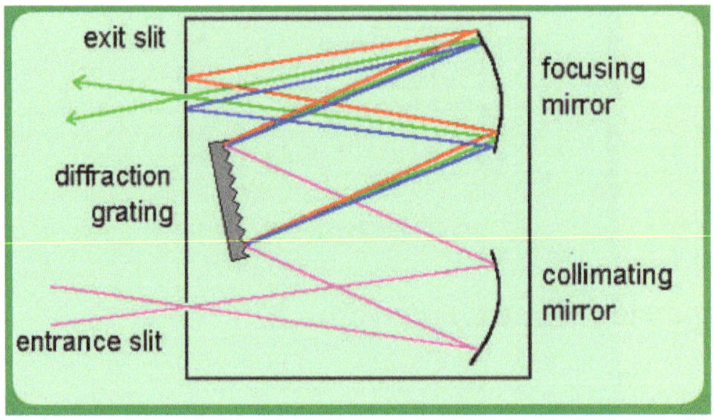

(i) Light gathering power / numerical aperture / speed of monochromator

Acceptance Angle = d/f_1

(ii) The spectral transmission of the optical systems $T(\lambda)$ or $R(\lambda)$

(iii) Spectral resolving power = $l/\Delta\lambda$

minimum separation $\Delta\lambda$ of two spectral lines that can be resolved

(iv) Free Spectral range : wavelength range $\Delta\lambda$ in which the wavelength λ can be unambiguously determined from $x(\lambda)$ =

If N is the no. of grooves per inch and m is the order then $\frac{\lambda}{\Delta\lambda} = mN$

References

- Littlejohn, Robert (Fall 2007). "Gaussian, SI and Other Systems of Units in Electromagnetic Theory" (PDF). Physics 221A, University of California, Berkeley lecture notes. Retrieved 2008-05-06
- Kimball Milton; J. Schwinger (18 June 2006). Electromagnetic Radiation: Variational Methods, Waveguides and Accelerators. Springer Science & Business Media. ISBN 978-3-540-29306-4
- J Rosen. "Redundancy and superfluity for electromagnetic fields and potentials". American Journal of Physics. 48 (12): 1071. Bibcode:1980AmJPh..48.1071R. doi:10.1119/1.12289
- Thomas B. A. Senior & John Leonidas Volakis (1995-03-01). Approximate Boundary Conditions in Electromagnetics. London UK: Institution of Electrical Engineers. p. 261 ff. ISBN 0-85296-849-3
- Narayanan, DL; Saladi, RN; Fox, JL (September 2010). "Ultraviolet radiation and skin cancer". International Journal of Dermatology. 49 (9): 978–86. doi:10.1111/j.1365-4632.2010.04474.x. PMID 20883261
- Kinsler, P. (2010). "Optical pulse propagation with minimal approximations". Phys. Rev. A. 81: 013819. arXiv:0810.5689. Bibcode:2010PhRvA..81a3819K. doi:10.1103/PhysRevA.81.013819
- Buchwald, Jed (1989). The Rise of the Wave Theory of Light: Optical Theory and Experiment in the Early Nineteenth Century. Chicago: University of Chicago Press. ISBN 0-226-07886-8. OCLC 18069573
- Saladi, RN; Persaud, AN (January 2005). "The causes of skin cancer: a comprehensive review". Drugs of today (Barcelona, Spain : 1998). 41 (1): 37–53. doi:10.1358/dot.2005.41.1.875777. PMID 15753968
- Binhi, Vladimir N (2002). Magnetobiology: Underlying Physical Problems. Repiev, A & Edelev, M (translators from Russian). San Diego: Academic Press. pp. 1–16. ISBN 978-0-12-100071-4. OCLC 49700531
- Carmichael, H. J. "Einstein and the Photoelectric Effect" (PDF). Quantum Optics Theory Group, University of Auckland. Archived from the original (PDF) on 17 October 2012. Retrieved 22 December 2009

- Kumar, Manjit (2011). Quantum: Einstein, Bohr, and the Great Debate about the Nature of Reality (Reprint ed.). W. W. Norton & Company. pp. 242, 375–376. ISBN 978-0393339888
- Schrödinger, Erwin (December 1926). “An Undulatory Theory of the Mechanics of Atoms and Molecules”. Phys. Rev. 28 (6): 1049–1070. Bibcode:1926PhRv...28.1049S. doi:10.1103/PhysRev.28.1049
- Teschl, Gerald (2009). Mathematical Methods in Quantum Mechanics; With Applications to Schrödinger Operators. Providence: American Mathematical Society. ISBN 978-0-8218-4660-5
- “Observing the quantum behavior of light in an undergraduate laboratory”. American Journal of Physics. 72: 1210. Bibcode:2004AmJPh..72.1210T. doi:10.1119/1.1737397
- Clauser, John F.; S. Li (1994). “Talbot von Lau interefometry with cold slow potassium atoms.”. Phys. Rev. A. 49 (4): R2213–17. Bibcode:1994PhRvA..49.2213C. doi:10.1103/PhysRevA.49.R2213. PMID 9910609

3

Atomic Structure of Hydrogen Atom

Equating a hydrogen-like atom with the Schrödinger equation is an important experiment. A hydrogen-like atom has a nucleus and one electron. The major components of the atomic structure of hydrogen are discussed in this chapter.

Angular Part of the Hamiltonian

- In this lesson, we will apply the Schrodinger Equation to the simple system "Hydrogen" and compare the outcome of the Quantum Theory with the experimental results obtained from different experiments.
- We will also get to know about the "wavefunction" and its use in developing the concept of the structure of atoms.

Hydrogen

- One electron atom and the simplest bound system in nature.
- Referring to the figure below, the proton (+ve Charge) is at the center and the electron (-ve charge) is moving around the center.
- The attractive force between them is Coloumbic in nature and the potential is:

$$V = V(x, y, z) = \frac{-Ze^2}{4\pi\varepsilon_0\sqrt{x^2 + y^2 + z^2}}$$

Where, is the charge of the nucleus, for hydrogen.

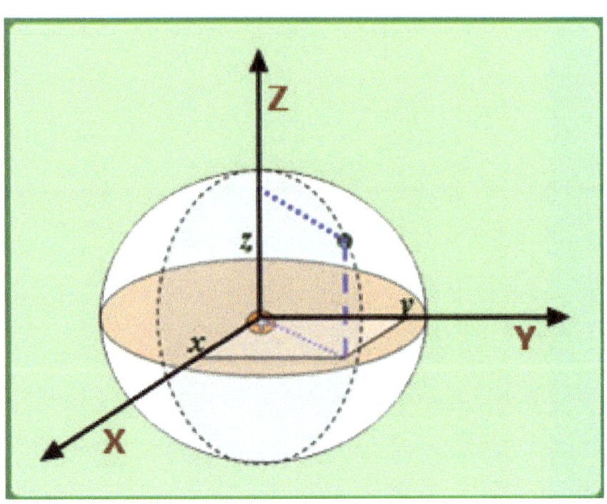

- Since it is a two body problem, we convert it to one body problem by introducing reduced mass:

$$\mu = \frac{mM}{m+M}$$

Where, m and M are the masses of the electron and proton, respectively.

- So, the total energy of the system = kinetic energy + potential energy

$$\frac{1}{2\mu}(p_x^2 + p_y^2 + p_z^2) + V(x, y, z) = E$$

Here, p_x , p_y and p_z are the linear momenta and E is the total energy.

$$p_x \rightarrow -i\hbar\frac{\partial}{\partial x}, p_y \rightarrow -i\hbar\frac{\partial}{\partial y}, p_z \rightarrow -i\hbar\frac{\partial}{\partial z}$$

- For the quantum mechanical treatment, we will convert the classical dynamical quantities (p_x , p_y and p_z) to its corresponding quantum mechanical operators.

$$p_x \rightarrow -i\hbar\frac{\partial}{\partial x}, p_y \rightarrow -i\hbar\frac{\partial}{\partial y}, p_z \rightarrow -i\hbar\frac{\partial}{\partial z}$$

$$E \rightarrow -i\hbar\frac{\partial}{\partial t}$$

- Substituting in Equation above, we get

$$-\frac{\hbar^2}{2\mu}\left(\frac{\partial^2}{\partial x^2} + \frac{\partial^2}{\partial y^2} + \frac{\partial^2}{\partial z^2}\right) + V(x, y, z) = i\hbar\frac{\partial}{\partial t}$$

- class=”objectivelist”>Here, we introduce the Wavefuntion to represent the electron.

$$\Psi = \Psi(x, y, z, t)$$

The wavefunction contains the information about the position and the time evolution of the electronic motion. We will understand more about this later.

So operating Equation- 2, on the wavefunction $\Psi = \Psi(x, y, z, t)$,

$$-\frac{\hbar}{2\mu}\left(\frac{\partial^2\Psi(x,y,z,t)}{\partial x^2} + \frac{\partial^2\Psi(x,y,z,t)}{\partial^2 y} + \frac{\partial^2\Psi(x,y,z,t)}{\partial z^2}\right) + V(x,y,z)\Psi(x,y,z,t) = i\hbar\frac{\partial\Psi(x,y,z,t)}{\partial t}$$

$$\Rightarrow -\frac{\hbar^2}{2\mu}\nabla^2\Psi + V\Psi = i\hbar\frac{\partial\Psi}{\partial t}$$

$$\nabla^2 = \frac{\partial^2}{\partial x^2} + \frac{\partial^2}{\partial y^2} + \frac{\partial^2}{\partial z^2} \quad \textit{is the Laplacian operator}$$

Since the potential $V(x,y,z)$ does not depend on time t, and we are interested to evaluate the energy of the stationary (time independent) states, we can take this wavefunction as the explicit dependence of time such as:

$$\Psi(x,y,z,t) = \Psi(x,y,z)e^{-iEt/\hbar}$$

Substituting this to Equation above, we get

$$-\frac{\hbar^2}{2\mu}\nabla^2\psi(x,y,z) + V\psi(x,y,z) = E\psi(x,y,z)$$

This is the Time Independent Schrodinger Equation.

Now we have to evaluate the stationary state energies (E) of the electron in this system by solving this equation to explain the observed spectra of hydrogen.

Since the potential is spherically symmetric

$$V(x,y,z) = \frac{Ze^2}{4\pi\varepsilon_0\sqrt{x^2+y^2+z^2}} = \frac{Ze^2}{4\pi\varepsilon_0 r} = V(r)$$

We can convert Equation above to its spherical polar coordinate form,

$$-\frac{\hbar^2}{2\mu}\nabla^2\psi(r,\theta,\phi) + V(r)\psi(r,\theta,\phi) = E\psi(r,\theta,\phi)$$

Radial distance $\Rightarrow r$, Polar angle $\Rightarrow \theta$ and Azimuthal angle $\Rightarrow \phi$

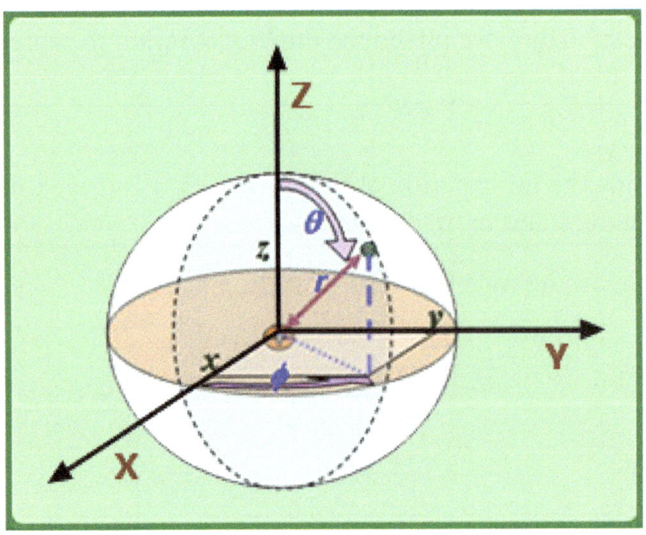

The form of the Laplacian operator in Spherical polar coordinates is:

$$\nabla^2 = \frac{1}{r^2}\frac{\partial}{\partial r}\left({}^2 \frac{\partial}{\partial r}\right) + \frac{1}{r^2 \sin\theta}\frac{\partial}{\partial\theta}\left(\sin \frac{\partial}{\partial\theta}\right) + \frac{1}{r^2 \sin^2\theta}\frac{\partial}{\partial\phi^2}$$

Using the separation of the variables, we can write the wavefunction as the product form of the independent variables $R(r), \Theta(\theta), \Phi(\phi)$

$$\psi(r,\theta,\phi) = R(r)\Theta(\theta)\Phi(\phi)$$

Substituting in Equation above,

$$-\frac{\hbar^2}{2\mu}\left[\frac{1}{r^2}\frac{\partial}{\partial r}\left(r^2\frac{\partial R\Theta\Phi}{\partial r}\right) + \frac{1}{r^2\sin\theta}\frac{\partial}{\partial\theta}\left(\sin\theta\frac{\partial R\Theta\Phi}{\partial\theta} + \frac{1}{r^2\sin^2\theta}\frac{\partial^2 R\Theta\Phi}{\partial\phi^2}\right)\right] + V(r)R\Theta\Phi = ER\Theta\Phi$$

Carrying out the partial differentiation,

$$-\frac{\hbar^2}{2\mu}\left[\frac{\Theta\Phi}{r^2}\frac{\partial}{\partial r}\left(r^2\frac{\partial R}{\partial r}\right) + \frac{R\Phi}{r^2\sin\theta}\frac{\partial}{\partial\theta}\left(\sin\theta\frac{\partial\Theta}{\partial\theta}\right) + \frac{R\Theta}{r^2\sin^2\theta}\frac{\partial^2\Phi}{\partial\phi^2}\right] + V(r)R\Theta\Phi = ER\Theta\Phi$$

Multiplying the this equation by $\frac{-2\mu r^2}{R\Theta\Phi\hbar^2}$ and rearranging we get,

$$\left[\frac{1}{R}\frac{\partial}{\partial r}\left(r^2\frac{\partial R}{\partial r}\right) + \frac{1}{\Theta\sin\theta}\frac{\partial}{\partial\theta}\left(\sin\theta\frac{\partial\Theta}{\partial\theta}\right) + \frac{1}{\Phi\sin^2\theta}\frac{\partial^2\Phi}{\partial\phi^2}\right] + \frac{2\mu r^2}{\hbar^2}[E - V(r)] = 0$$

Separating the radial ($R(r)$) and angular ($\Theta(\theta), \Phi(\phi)$) part:

$$\frac{1}{R}\frac{d}{dr}\left(r^2\frac{dR}{dr}\right) + \frac{2\mu r^2}{\hbar^2}[E - V(r)] = -\frac{1}{\Theta\sin\theta}\frac{d}{d\theta}\left(\sin\theta\frac{d\Theta}{d\theta}\right) - \frac{1}{\Phi\sin^2\theta}\frac{d^2\Phi}{d\phi^2}$$

Note here that the partial derivative forms are converted to total derivative form. Now each equation should be equal to a constant, let's take as λ.

So, the radial part is:

$$\frac{1}{R}\frac{d}{dr}\left(r^2\frac{dR}{dr}\right) + \frac{2\mu r^2}{\hbar^2}[E - V(r)] = \lambda$$

$$\frac{1}{r^2}\frac{d}{dr}\left(r^2\frac{dR}{dr}\right) + \left[\frac{2\mu}{\hbar^2}(E - V(r)) - \frac{\lambda}{r^2}\right]R = 0$$

And the angular part is:

$$\frac{1}{\Theta \sin\theta}\frac{d}{d\theta}\left(\sin\theta\frac{d\Theta}{d\theta}\right)+\frac{1}{\Phi\sin^2\theta}\frac{d^2\Phi}{d\phi^2}=-\lambda$$

$$\Rightarrow\frac{\sin\theta}{\Theta}\frac{\partial}{d\theta}\left(\sin\theta\frac{d\Theta}{d\theta}\right)+\frac{1}{\Phi}\frac{d^2\Phi}{d\phi^2}=-\lambda\sin^2\theta$$

Now, separating the Polar and the Azimuthal part:

$$\frac{\sin\theta}{\Theta}\frac{d}{d\theta}\left(\sin\theta\frac{d\Theta}{d\theta}\right)+\lambda\sin^2\theta=-\frac{1}{\Phi}\frac{d^2\Phi}{d\phi^2}=m_i^2$$

Here, we have taken that both sides should be equal to a constant m^2. Note that we have three independent equations from Equations above.

Now let us first consider the azimuthal part,

$$-\frac{1}{\Phi}\frac{d^2\Phi}{d\phi^2}=m_i^2$$

$$\frac{1}{\Phi}\frac{d^2\Phi}{d\phi^2}+m_i^2\Phi=0$$

The general solution of this equation may be written in the form:

$$\Phi_{ml}(\phi)=\frac{1}{\sqrt{2\pi}}r^{im_l\phi}$$

Which satisfies the orthonormality condition $\int_0^{2\pi}\Phi^*_{m_l}(\phi)\Phi_{m_l'}(\phi)d\phi=\delta_{m_lm_l'}$

As the eigen functions must be single valued, i.e., $\Phi(0)=\Phi(2\pi)$ which means

$$e^{im_l0}=e^{im_l2\pi}$$

And using Euler's formula $1=\cos(m_l2\pi)+i\sin(m_l2\pi)$

This is satisfied only if $m_l=0,\pm1,\pm2,.....$

Therefore, acceptable solutions only exist when m_l can only have certain integer values, i.e. it is a quantum number. Thus, m_l is called the magnetic quantum number in spectroscopy because it plays role when atom interacts with magnetic fields.

$$\frac{\sin\theta}{\Theta}\frac{d}{d\theta}\left(\sin\theta\frac{d\Theta}{d\theta}\right)+\lambda\sin^2\theta=m_l^2$$

Rearranging, we get

$$\frac{1}{\sin\theta}\frac{d}{d\theta}\left(\sin\theta\frac{d\Theta(\theta)}{d\theta}\right)+\left(\lambda-\frac{m_l^2}{\sin^2\theta}\right)\Theta(\theta)=0$$

Let us introduce new variable $\mu=\cos\theta$, the Equation above becomes,

$$\frac{d}{d\omega}\left((1-\omega^2\frac{d\Theta(\omega)}{d\omega}\right)+\left(\lambda-\frac{m_l^2}{1-\omega^2}\right)\Theta(\omega)=0$$

The equation above has singularities at $\omega=\pm1$, which may be eliminated by having a solution Θ in the form of

$$\Theta=(1-\omega^2)^{s/2}u$$

Then equation above becomes

$$(1-\omega^2)\frac{d^2u}{d\omega^2}-2\omega(s+1)\frac{du}{d\omega}+\left[\lambda+\frac{s^2\omega^2-m_l^2}{1-\omega^2}\right]u=0$$

Which can be written as:

$$(1-\omega^2)\frac{d^2u}{d\omega}-2\omega(s+1)\frac{du}{d\omega}+\left[\lambda-s^2-s+\frac{s^2-m_l^2}{1-\omega^2}\right]u=0$$

The last singular term in Equation above can be removed by taking $s=\pm m_l\geq 0$, and then we have

$$(1-\omega^2)\frac{d^2u}{d\omega^2}-2\omega(m_l+1)\frac{du}{d\omega}+[\lambda-m_l(m_l+1)]u=0$$

Which is a regular equation and hence its series solution may be written as:

$$u=\sum_{r-0} a_r\omega^r$$

Substituting in Equation above, yields the recursion relation:

$$a_{r+2}=\frac{(r+m_l)(r+m_l+1)-\lambda}{(r+1)(r+2)}a_r$$

Requiring that the series be (Equation above) be limited by a certain power $r=q$, i.e. requiring that it be a polynomial of order q, we have to introduce the condition

$a_{r+2}=0, a_r\neq 0$ which requires $\lambda=(q+m_l)(q+m_l+1)$

Where $q=0,1,2,3,\ldots\ldots$

Now we introduce $l=q+m_l=0,1,2,3,\ldots\ldots.$ (orbital quantum number) such that

$l \geq m_l$

$\lambda = l(l+1)$

Then Equation becomes:

$$\frac{1}{\sin\theta}\frac{d}{d\theta}\left(\sin\theta\frac{d\Theta(\theta)}{d\theta}\right)+\left(l(l+1)-\frac{m_1^2}{\sin^2\theta}\right)\Theta(\theta)=0$$

This is well known associated Legendre differential equation with its solution as associated Legendre polynomials.

$$\Theta_{l,m_l} = C_l^{m_l} P_l^{m_l}(\omega)$$

Where $C_l^{m_l}$ is the normalization factor and $P_l^{m_l}(\omega)$ is the associated Legendre polynomial defined by

$$P_l^{m_l} = (1-\omega^2)^{m_l/2}\frac{d^{l+m_l}}{d\omega^{l+m_l}}\left[\frac{(\omega^2-1)^l}{2^l l!}\right]$$

$$= (1-\omega^2)^{m_l/2}\frac{d^{m_l}}{d\omega^{m_l}}P_l(\omega)$$

Where $p_l(\omega)$ is the ordinary Legendre Polynomial. This expression holds for negative values of m_l also

$$p_l^{m_l}(\omega) = (-1)^{m_l}\frac{(l+m_l)!}{(l-m_l)!}p_l^{-m_l}(\omega)$$

From this we can establish the range of variation of the azimuthal (magnetic) quantum number

$$m_l = 0, \pm1, \pm2, \pm3, \ldots\ldots, \pm l$$

The associated Legendre Polynomials satisfy the orthonormalization property,

$$\int_{-1}^{1} P_l^{m_l}(\omega)P_l^{m_l}(\omega)d\omega = \frac{2}{2l+1}\frac{(l+m_l)!}{(l-m_l)!}\delta_{l_v}v$$

So,

$$C_l^{m_l} = \sqrt{\frac{2l+1}{2}\frac{(l-m_1)!}{(l+m_l)!}}$$

Then Equation above becomes

$$\Theta_{l,m_l} = \sqrt{\frac{2l+1}{2}\frac{(l-m_l)!}{(1+m_l)!}}\,P_l^{m_l}(\omega)$$

So, we have now two quantum numbers, namely, Orbital (l) and magnetic (m_l)

$$l = 0,1,2,3,.....$$
$$m_l = -l,-l+1,.....,0,....,l-1,l$$

Depending on the value of l and m_l, some of the Associated Legendre polynomials are:

$p_0^0 = 1$

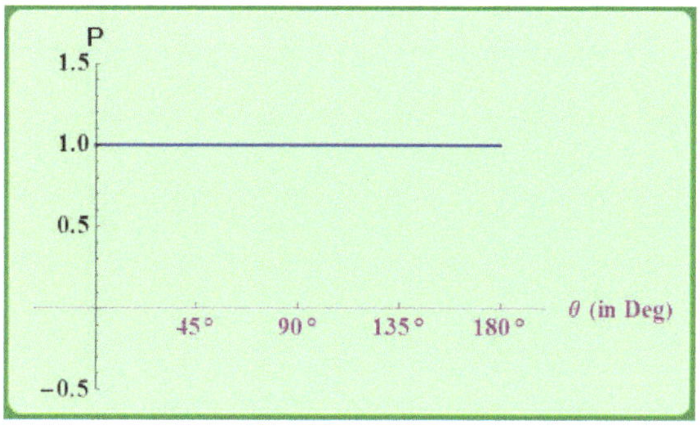

$p_1^0 = \cos\theta$

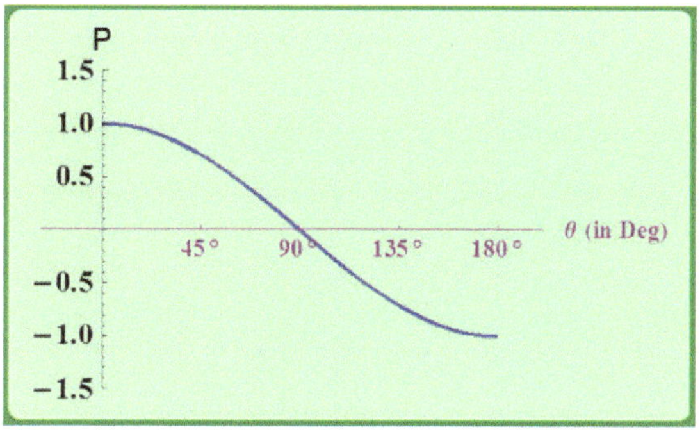

$p_1^{\pm 1} = (1-\cos^2\theta)^{1/2}$

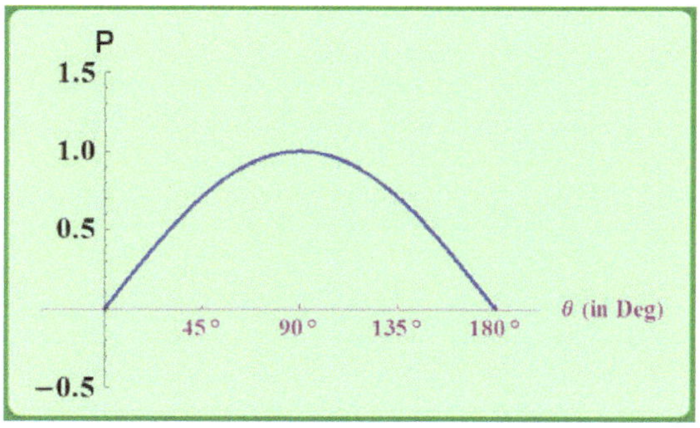

$p_2^0 = 1-3\cos^2\theta$

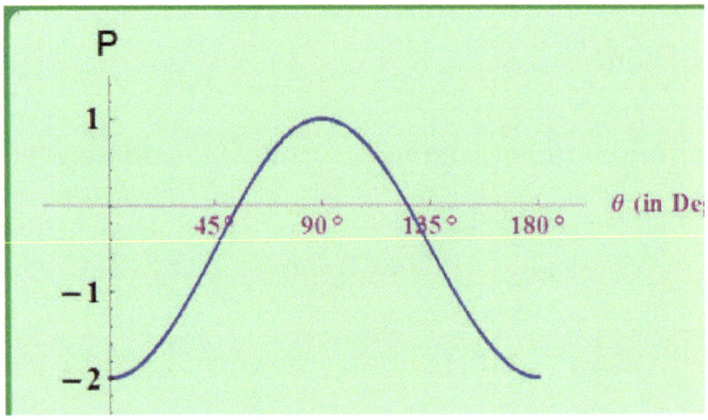

$p_2^{\pm 1} = (1-\cos^2\theta)^{1/2}\cos\theta$

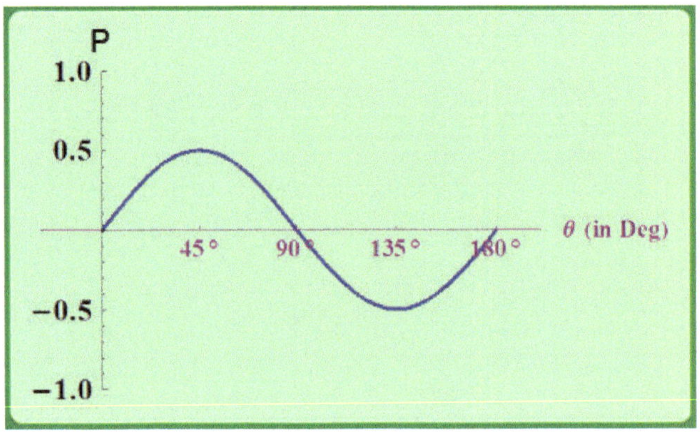

$P_2^{\pm 2} = 1-\cos^2\theta$

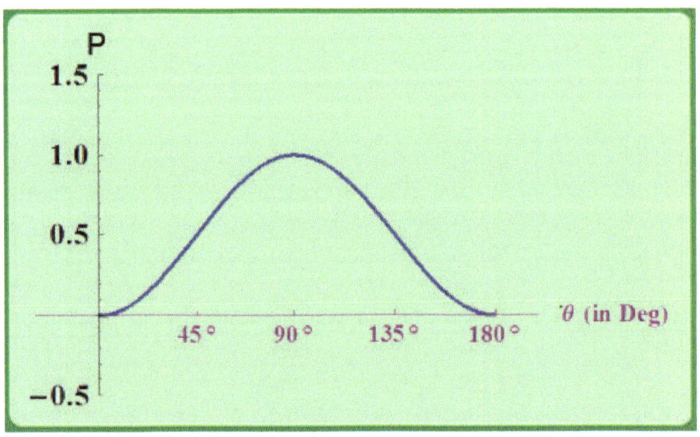

Now, we can have the total angular part, from Equations above

$$Y_l^{m_l}(\theta,\phi) = \Theta_{l,m_l}(\theta)\Phi_{m_l}(\phi)$$

$$= \sqrt{\frac{2l+1}{4\pi}\frac{(l-m_l)!}{(l+m_l)!}} P_l^{m_l}(\cos\theta)e^{im_l\phi}$$

$$|Y_0^0|^2 = |1|^2$$

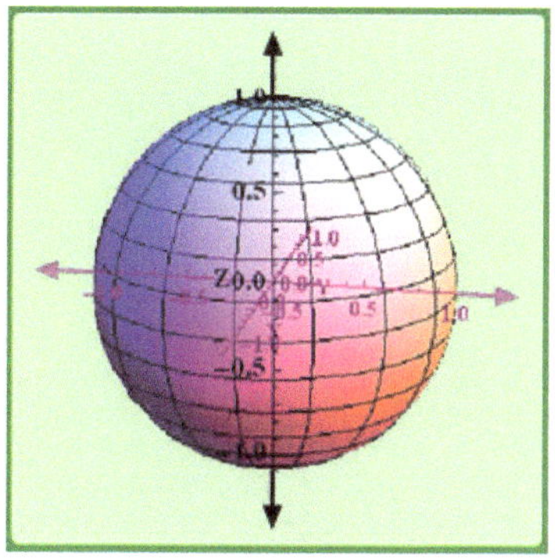

$$|Y_1^0|^2 = |\cos\theta|^2$$

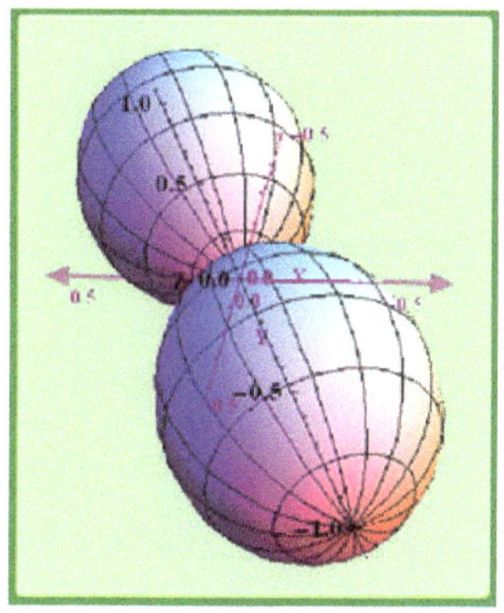

Radial Part of the Hamiltonian

Now, let us solve the radial part

$$\frac{1}{r^2}\frac{d}{dr}\left(r^2\frac{dR}{dr}\right)+\left[\frac{2\mu}{\hbar^2}\left(E+\frac{Ze^2}{4\pi\varepsilon_0 r}\right)-\frac{\ell(\ell+1)}{r^2}\right]R=0$$

Let us first evaluate the ground state (lowest energy) of hydrogen atom.

Assuming the ground state and taking $\ell = 0$, we get

$$\frac{1}{r^2}\frac{d}{dr}\left(r^2\frac{dR}{dr}\right)+\frac{2\mu}{\hbar^2}\left(E+\frac{Ze^2}{4\pi\varepsilon_0 r}\right)R=0$$

$$\Rightarrow \frac{1}{r^2}\left[r^2\frac{d^2R}{dr^2}+2r\frac{dR}{dr}\right]+\frac{2\mu}{\hbar^2}\left(E+\frac{Ze^2}{4\pi\varepsilon_0 r}\right)R=0$$

$$\Rightarrow \frac{d^2R}{dr^2}+\frac{2}{r}\frac{dR}{dr}+\frac{2\mu}{\hbar^2}\left(E+\frac{Ze^2}{4\pi\varepsilon_0 r}\right)R=0$$

Let us try a solution $R(r) = Ae^{-r/a_0}$, where A and a_0 are constants.

$$\frac{dR(r)}{dr} = -\frac{A}{a_0}e^{-r/a_0}; \frac{d^2R}{dr} = \frac{A}{a_0^2}e^{-r/a_0}$$

$$\frac{A}{a_0^2}e^{-r/a_0} - \frac{2A}{ra_0}e^{-r/a_0} + \frac{2A\mu}{\hbar^2}\left(E + \frac{Ze^2}{4\pi\varepsilon_0 r}\right)e^{-r/a_0} = 0$$

$$\Rightarrow \left(\frac{1}{a_0^2} + \frac{2\mu E}{\hbar^2}\right) + \left(\frac{2\mu Ze^2}{4\pi\varepsilon_0\hbar^2} - \frac{2}{a_0}\right)\frac{1}{r} = 0$$

To satisfy this equation for any value of *r*,

$$\frac{1}{a_0^2} + \frac{2\mu E}{\hbar^2} = 0$$

$$\Rightarrow E = -\frac{\hbar^2}{2\mu a_0^2}$$

and

$$\frac{2\mu Ze^2}{4\pi\varepsilon_0\hbar^2} - \frac{2}{a_0} = 0$$

$$\Rightarrow a_0 = \frac{4\pi\varepsilon_0\hbar^2}{\mu Ze^2} = \frac{4\pi\varepsilon_0\hbar^2}{\mu e^2} \qquad [for\ Z = 1]$$

Let us calculate the values of a_0 and Ground State Energy *E*,

$$\mu = \frac{m_e m_p}{m_e + m_p}$$

$$\approx 9.109 \times 10^{-31} Kg$$

$$\begin{pmatrix} where\ m_e = 9.109 \times 10^{-31} Kg \\ m_p = 1.672 \times 10^{-27} Kg \\ and\ \ m_p >> m_e \end{pmatrix}$$

$$e = 1.6 \times 10^{-19} Coul$$

$$\frac{1}{4\pi\varepsilon_0} = 8.988 \times 10^{9} {nt\text{-}m^2}/{Coul^2}$$

$$\hbar = 1.055 \times 10^{-34} J - \sec = 0.6582 \times 10^{-15} eV - \sec$$

$$a_0 = \frac{(1.055 \times 10^{-34})^2}{8.988 \times 10^9 \times (1.6 \times 10^{-19})^2 \times 9.109 \times 10^{-31}}$$

$$= \frac{1.055 \times 1.055}{8.988 \times 1.6 \times 1.6 \times 9.109} \times 10^{-34-34+19+19+31-9}$$

$$= 0.00531 \times 10^{-8} m = 0.529 \times 10^{-10} m = 0.529 \mathring{A}$$

which is same as Bohr Radius.

Substituting value of a_0 in Eqn. above,

$$E = -\frac{\hbar^2}{2\mu a_0^2}$$

$$= \frac{(1.055\times10^{-34})^2}{2\times9.109\times10^{-31}\times(0.529\times10^{-10})^2} = \frac{1.113\times10^{-68}}{5.098\times10^{-51}}$$

$$= \frac{0.2183\times10^{-17}}{1.6\times10^{-19}} = 0.1364\times10^{+2} = -13.6eV$$

This is the lowest energy state of Hydrogen obtained also from Bohr's calculation.

Now let us calculate the general radial equation,

$$\frac{1}{r^2}\frac{d}{dr}\left(r^2\frac{dR}{dr}\right) + \left[\frac{2\mu}{\hbar^2}\left(E + \frac{Ze^2}{4\pi\varepsilon_0 r}\right) - \frac{\ell(\ell+1)}{r^2}\right]R = 0$$

The radial derivatives simplify if one factor $1/r$ out from function R, taking

$$R(r) = \frac{u(r)}{r}$$

$$\frac{dR}{dr} = \frac{r\dfrac{du}{dr} - u}{r^2}$$

$$\Rightarrow r^2\left(\frac{dR}{dr}\right) = r\frac{du}{dr} - u$$

$$\Rightarrow \frac{d}{dr}\left(r^2\frac{dR}{dr}\right) = r\frac{d^2u}{dr^2} + \frac{du}{dr} - \frac{du}{dr}$$

$$\Rightarrow \frac{1}{r^2}\frac{d}{dr}\left(r^2\frac{dR}{dr}\right) = r\frac{d^2u}{dr^2}$$

Substituting,

$$\frac{1}{r}\frac{d^2u(r)}{dr^2} + \left[\frac{2\mu}{\hbar}\left(E + \frac{Ze^2}{4\pi\varepsilon_0 r}\right) - \frac{\ell(\ell+1)}{r^2}\right]\frac{u(r)}{r} = 0$$

$$\Rightarrow \frac{d^2u(r)}{dr^2} + \left[\frac{2\mu}{\hbar}\left(E + \frac{Ze^2}{4\pi\varepsilon_0 r}\right) - \frac{\ell(\ell+1)}{r^2}\right]u(r) = 0$$

This is the Schrödinger equation for the particle in one dimension, restricted to $r > 0$.

Rearranging

$$\frac{d^2u(r)}{dr^2} + \left[\frac{2\mu E}{\hbar^2} - \underbrace{\left\{-\frac{2\mu Ze^2}{4\pi\varepsilon_0\hbar^2 r} + \frac{\ell(\ell+1)}{r^2}\right\}}_{V_{effects}}\right]u(r) = 0$$

This is a kind of potential $(for\,\ell \neq 0)$ with positive infinity at the origin, then negative potential and going to zero at large distances. So the minimum of the potential is at some positive r.

For bound states of proton-electron system, E will be a negative quantity.

Now, we will simplify by introducing dimensionless variable, ρ such as

$$\rho = +\alpha r$$
$$d\rho = +\alpha dr$$
$$d\rho^2 = \alpha^2 dr^2$$

$$\frac{\hbar^2}{2\mu}\frac{d^2u(r)}{dr^2} + \left[E + \frac{Ze^2}{4\pi\varepsilon_0 r} - \frac{\hbar^2}{2\mu}\frac{\ell(\ell+1)}{r^2}\right]u(r) = 0$$

$$\Rightarrow \frac{\hbar^2}{2\mu}\alpha^2\frac{d^2u(\rho)}{d\rho^2} + \left[E + \frac{Ze^2\alpha}{4\pi\varepsilon_0\rho} - \frac{\hbar^2}{2\mu}\frac{\alpha^2\ell(\ell+1)}{\rho^2}\right]u(\rho) = 0$$

$$\Rightarrow -\frac{\hbar^2}{2\mu}\frac{2\mu E}{\hbar^2}\frac{d^2u(\rho)}{d\rho^2} + \left[E + \frac{Ze^2\alpha}{4\pi\varepsilon_0\rho} - \frac{\hbar^2}{2\mu}\frac{2\mu E}{\hbar^2}\frac{\ell(\ell+1)}{\rho^2}\right]u(\rho) = 0$$

$$\Rightarrow -\frac{d^2u(\rho)}{d\rho^2} + \left[1 + \frac{Ze^2\alpha}{4\pi\varepsilon_0 E}\frac{1}{\rho} + \frac{\ell(\ell+1)}{\rho^2}\right]u(\rho) = 0$$

$$\Rightarrow \frac{d^2u(\rho)}{d\rho^2} = \left[1 - \frac{Ze^2\alpha}{4\pi\varepsilon_0 E}\frac{2\mu E}{\hbar^2\alpha}\frac{1}{\rho} + \frac{\ell(\ell+1)}{\rho^2}\right]u(\rho)$$

$$= \left[1 - \frac{Z\mu e^2}{2\pi\varepsilon_0\hbar^2\alpha}\frac{1}{\rho} + \frac{\ell(\ell+1)}{\rho^2}\right]u(\rho)$$

Substituting $\rho_0 = \dfrac{Z\mu e^2}{2\pi\varepsilon_0\hbar^2\alpha}$, we get

$$\frac{d^2u(\rho)}{dp^2} = \left[1 - \frac{\rho_0}{\rho} + \frac{\ell(\ell+1)}{\rho^2}\right]u(\rho)$$

Here, we understand the asymptotic behavior of solution of the earlier equation.

Case-I: when ρ is very large (at large separation between proton and electron), we can neglect the ρ terms in earlier equation and we get,

$$\frac{d^2u(p)}{dp^2}=u(\rho)$$

The solution is of the type,

$$u(\rho)=Ae^{-\rho}+Be^{+\rho}$$

Since e^{ρ} becomes infinite for large , we take and hence,

$$u(\rho)\sim Ae^{-\rho}$$

Case-III: When ρ is very small, then $\frac{1}{\rho^2}$ is the dominant term, so,

$$\frac{d^2u(\rho)}{d\rho^2}=\frac{\ell(\ell+1)}{\rho^2}u(p) \quad [note\ \ell\neq 0]$$

The solution, $\frac{d^2u(\rho)}{d\rho^2}=C\rho^{\ell+1}+D\rho^{-\ell}$

Again, as $\rho\to 0$; $\rho^{-\ell}$ becomes infinite.

So we take, $D=0$

So, for small ρ,

$$u(\rho)\sim C\rho^{\ell+1}$$

So, we have established that the solution $u(\rho)$ decays as $e^{-\rho}$ at large distance and goes as $\rho^{\ell+1}$ close to the proton (origin).

Now we write the general solution of equation as

$$u(\rho)=\rho^{\ell+1}e^{-\rho}\omega(\rho)$$

where, $\omega(\rho)$ we have to define

Let us substitute $u(\rho)$ into the Eqn. above

$$\frac{du(\rho)}{d\rho}=\rho^{\ell+1}e^{-\rho}\frac{d\omega(\rho)}{d\rho}+(\ell+1)\rho^{\ell}e^{-\rho}\omega(\rho)-\rho^{\ell+1}e^{-\rho}\omega(\rho)$$

$$=\rho^{\ell}e^{-\rho}\left[\rho\frac{d\omega(\rho)}{d\rho}+(\ell+1-\rho)\omega(\rho)\right]$$

and

$$\frac{d^2u(\rho)}{d\rho} = e^{-\rho}\rho^{\ell}\left[\rho\frac{d^2\omega(\rho)}{d\rho^2} + 2(\ell+1-\rho)\frac{d\omega(\rho)}{d\rho} + \left\{\frac{\ell(\ell+1)}{\rho} - 2(\ell+1) + \rho\right\}\omega(\rho)\right]$$

Substituting in Eqn. above

$$\cancel{e^{-\rho}\rho^{\ell}}\left[\rho\frac{d^2\omega(\rho)}{d\rho^2} + 2(\ell+1-\rho)\frac{d\omega(\rho)}{d\rho} + \left\{\frac{\ell(\ell+1)}{\rho} - 2(\ell+1) + \rho\right\}\omega(\rho)\right]$$

$$= \rho\left[1 - \frac{\rho_0}{\rho} + \frac{\ell(\ell+1)}{\rho^2}\right]\cancel{e^{-\rho}\rho^{1}}\omega(\rho)$$

$$\Rightarrow \rho\frac{d^2\omega(\rho)}{d\rho^2} + 2(\ell+1-\rho)\frac{d\omega(\rho)}{d\rho} + \left\{\cancel{\frac{\ell(\ell+1)}{\rho}} - 2(\ell+1) + \cancel{\rho} - \cancel{\rho} + \rho_0 - \cancel{\frac{\ell(\ell+1)}{\rho}}\right\}\omega(\rho) = 0$$

$$\Rightarrow \rho\frac{d^2\omega(\rho)}{d\rho^2} + 2(\ell+1-\rho)\frac{d\omega(\rho)}{d\rho} + \{\rho_0 - 2(\ell+1)\}\,\omega(\rho) = 0$$

Now, we have to solve this equation above, let us take a solution of the form,

$$\omega(\rho) = \sum_{j-0}^{\omega} C_j p^j$$

$$\frac{d\omega(\rho)}{d\rho} = \sum_{j-0}^{\omega} jC_j p^{j-1}$$

So,

$$\frac{d^2\omega(\rho)}{d\rho^2} = \sum_{j-0}^{\omega} j(j-1)C_j p^{j-2}$$

and

Putting back to Eqn. above

$$\sum_{j-0}^{\omega} j(j-1)C_j\rho^{j-1} + 2(\ell+1)\sum_{j-0}^{\omega} jC_j\rho^{j-1} - 2\sum_{j-0}^{\omega} jC_j\rho^j + (\rho_0 - 2(\ell+1))\sum_{j-0}^{\omega} C_j\rho^j = 0$$

Shifting the summation of the ρ^{j-1} terms

$$\sum_{j-1}^{\omega}(j+1)jC_{j+1}\rho^j + 2(\ell+1)\sum_{j-1}^{\omega}(j+1)C_{j+1}\rho^j - 2\sum_{j-0}^{\omega} jC_j\rho^j + (\rho_0 - 2(\ell+1))\sum_{j-0}^{\omega} C_j\rho^j = 0$$

For, $=-$, the first two terms becomes zero. So, we start from $j = 0$

$$\sum_{j-0}^{\omega} j(j+1)C_{j+1}\rho^j + 2(\ell+1)\sum_{j-0}^{\omega}(j+1)C_{j+1}\rho^j - 2\sum_{j-0}^{\omega} jC_j\rho^j + (\rho_0 - 2(\ell+1))\sum_{j-0}^{\omega} C_j\rho^j = 0$$

$$\Rightarrow \sum_{j-0}^{\omega}[j(j+1)C_{j+1} + 2(\ell+1)(j+1)C_{j+1} - 2jC_j + (\rho_0 - 2(\ell+1))C_j]\rho^j = 0$$

This is valid, if all the coefficients of ρ^j are zero.

$$j(j+1)C_{j+1} + 2(\ell+1)(j+1)C_{j+1} - 2jC_j + (\rho_0 - 2(\ell+1))C_j = 0$$

$$\Rightarrow C_{j+1} = \frac{(2j + 2(\ell+1)) - \rho_0}{(j+1)(j+2(\ell+1))} C_j$$

Let us examine the behavior at large j value. For large j value, we ignore $(\ell+1)$ and ρ_0, so,

$$C_{j+1} = \frac{2}{(j+1)} C_j$$

Taking the value c_0 for $j = 0$

$$C_1 = \frac{2}{1} C_0$$

$$C_2 = \frac{2}{2} C_1 = \frac{2^2}{1\times 2} C_0$$

$$C_3 = \frac{2}{3} C_2 = \frac{2^3}{1\times 2\times 3} C_0$$

$$\therefore C_j = \frac{2^j}{j!} C_0$$

Thus,

$$\omega(\rho) = \sum_{j-0}^{\omega} C_j \rho^j = C_0 \sum_{j-0}^{\omega} \frac{2^j}{j!} \rho^j$$

$$= C_0 e^{2p}$$

Returning back in Eqn. above, the general solution,

$$u(\rho) = \rho^{\ell+1} e^{-\rho} \omega(\rho)$$

$$= \rho^{\ell+1} e^{-\rho} C_0 e^{2\rho}$$

$$= C_0 \rho^{\ell+1} e^{\rho}$$

Again, the value of $u(\rho)$ increases exponentially for large ρ and cannot be accepted. The series has to terminate at finite number before reaching the large value of ρ.

So, we should except, that at some finite value of j, Eqn. above

$$2(j + \ell + 1) = \rho_0$$

So, ρ_0 should be an even integer.

We define, $\rho_0 = 2n$

$$\frac{Z\mu e^2}{2\pi\varepsilon_0\hbar^2\alpha} = 2n$$

So, $\alpha = \dfrac{Z\mu e^2}{4\pi\varepsilon_0\hbar^2 n}$

and,

$$\sqrt{-\frac{2\mu E}{\hbar^2}} = \frac{Z\mu e^2}{4\pi\varepsilon_0\hbar^2 n}$$

$$\Rightarrow -\frac{2\mu E}{\hbar^2} = \frac{Z^2\mu^2 e^4}{(4\pi\varepsilon_0)^2\hbar^2 n^2}$$

$$\Rightarrow E = -\frac{Z^2\mu e^4}{(4\pi\varepsilon_0)^2 2\hbar^2}\frac{1}{n^2}$$

This is same energy as Bohr formula for Hydrogen putting $Z = 1$

Substituting $\rho_0 = 2n$ in Eqn. above

$$C_{j+1} = \frac{2(j+\ell+1)-2n}{(j+1)(j+2(\ell+1))}C_j$$

$$= -\frac{2(n-\ell-1-j)-2n}{(j+1)(2\ell+j+2)}C_j$$

In fact, the solution of Eqn. above is known as Associated Laguerre polynomials which is of the form

$$\omega(\rho) = L_{n-\ell-1}^{2\ell+1}(2\rho)$$

$$= \sum_{j-0}^{\omega}\frac{(-1)^j 2^j (n+\ell)!}{(n-\ell-j-1)!(2\ell+j+1)!j!}$$

So, the coefficients

$$C_j = \frac{(-1)^j 2j(n+\ell)!}{(n-\ell-j-1)!(2\ell+j+1)!j!}$$

$$\frac{C_{j+1}}{C_j} = -\frac{2(n-\ell-1-j)}{(j+1)(2\ell+j+2)}$$

which is same as before.

In fact, rearranging Eqn. above we can get the solution as “Associated Laguerre Polynomials” of the differential equation of the form,

$$Z\frac{d^2}{dZ^2}L_n^k(Z)+(k+1-Z)\frac{d}{dZ}L_n^k(Z)+\bar{n}L_n^k(Z)=0$$

Substituting $Z=2p$ in Eqn. above, we get,

$$Z\frac{d^2}{dZ^2}\omega(Z)+[(2\ell+1)+1-Z]\frac{d\omega(z)}{dZ}+\left[\frac{\rho_0}{2}-(\ell+1)\right]\omega(Z)=0$$

$$\bar{n}=\frac{\rho_0}{2}-(\ell+1)$$

$$Here, K=2\ell+1, and=\frac{2n}{2}-(\ell+1)=n-(\ell+1)$$

So the solution,

$$\omega(Z)=L_{n-(\ell+1)}^{2\ell+1}(Z)=L_{n-(\ell+1)}^{2\ell+1}(2p)$$

In this case, the solution above has to be rearranged and we get,

$$u(p)=(2p)^{\ell+1}e^{-p}L_{n-(\ell+1)}^{2\ell+1}(2p)$$

So,

$$rR_{n\ell}(r)=u(r)=\left(\frac{2r}{na_0}\right)^{\ell+1}e^{-r/na_0}L_{n-(\ell+1)}^{2\ell+1}\left(\frac{2r}{na_0}\right)$$

$$\Rightarrow R_{n,\ell}(r)=N_{n,\ell}e^{-r/na_0}\left(\frac{2r}{na_0}\right)^{\ell}L_{n-(\ell+1)}^{2\ell+1}\left(\frac{2r}{na_0}\right)$$

Here we have added the normalization constant $N_{n\ell}$ and absorbed the factor $2/na_0$ from the power term of r

Now, we evaluate the normalization constant $N_{n\ell}$ from the relation,

$$\int_0^\omega R_{n,\ell}^*(r)R_{n,\ell}(r)r^2dr=1$$

Substituting,

$$|N|^2\left(\frac{na_0}{2}\right)^3\int_0^\omega e^{-2r/na_0}\left(\frac{2r}{na_0}\right)^{2\ell+2}\left[L_{n-\ell-1}^{2\ell+1}\left(\frac{2r}{na_0}\right)\right]^2 d\left(\frac{2r}{na_0}\right)=1$$

Here, we can use the orthogonality relation,

$$\int_0^\omega e^{-z}Z^{k+1}L_n^k(z)L_m^k(z)dZ=\frac{\{(n+k)!\}^3}{n!}(2n+k+1)\partial mn$$

So, we get, $|N_{n,\ell}|^2 \left(\frac{na_0}{2}\right)^3 \frac{2n\left\{(n+\ell)!\right\}^3}{(n-\ell-1)!} = 1$

So, $N_{n,\ell} = \sqrt{\left(\frac{2}{na_0}\right)^3 \frac{(n-\ell-1)!}{2n\left\{(n+\ell)!\right\}^3}}$

Normalized Radial Wavefunction,

$$R_{n,\ell}(r) = \sqrt{\left(\frac{2}{na_0}\right)^3 \frac{(n-\ell-1)!}{2n\left\{(n+\ell)!\right\}^3}} e^{-r/na_0} \left(\frac{2}{na_0}\right)^\ell * L_{n-\ell-1}^{2\ell+1}\left(\frac{2}{na_0}\right)$$

Some useful Associated Laguerre Polynomials

$L_0^1(z) = 1$

$L_1^1(z) = 4 - 2Z$ $\quad L_0^3(Z) = 6$

$L_2^1(z) = 3Z^2 - 18Z + 18$ $\quad L_1^3(z) = 96 - 24Z$

Interpretation of Wavefunction and Modification for Alkali Atoms

- The wavefunction calculated for the hydrogen atom is complex quantity (includes $e \pm im_l\Phi$).
- This can not be measured by any actual physical instrument. However, it contains all the information which the uncertainty principle allows us to know about the associated electrons.
- The connection between the properties of the wavefunction $\Psi(x,t)$ and the behavior of the electron is expressed in terms of the probability density $p(x,t)$.
- $p(x,t)$ is the probability per unit length of finding the particle near the coordinate x at time t
- According to Max Born, 1962, the probability density $p(x,t) = \Psi^*(x,t)\Psi(x,t)$ where $\Psi^*(x,t)$ represents the complex conjugate of $\Psi(x,t)$.

Born Interpretation of the Wavefunction

If at any instant of time 't', a measurement is made to locate the electron with the wavefunction $\Psi(x,t)$ the probability $p(x,t)dx$ of finding a particle at a position between x and x+dx is proportional to $|\Psi(x,t)|^2 dx$

$$p(x,t) = \Psi^*(x,t)\Psi(x,t) = |\Psi(x,t)|^2$$

It implies that sign of the wavefunction has no direct physical significance. In figure (a), the wavefunction has changed its sign but the probability of finding the electron is symmetric with respect to the nucleus at r/ao = 0.

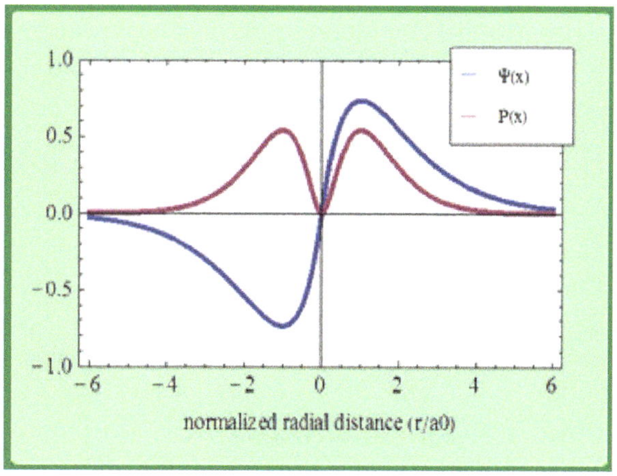

Figure (a)

If the wavefunction changes sign as shown in the figure (b), it is called node ($\Psi(x,t)=0$) The probability of finding the electron (Figure - c) at node is zero because

$$p(x,t)=\Psi^*(x,t)\Psi(x,t)=|\Psi(x,t)|^2$$

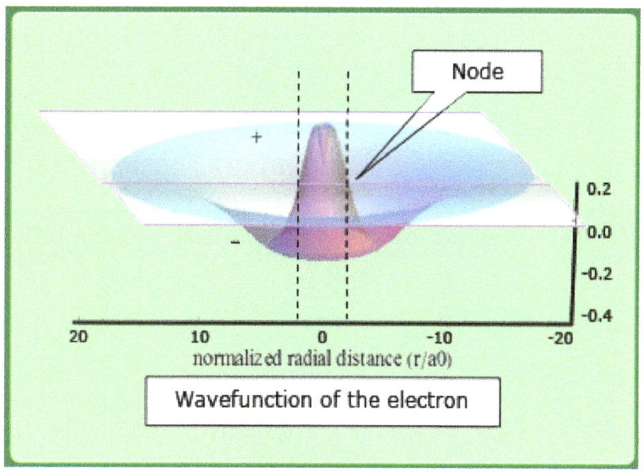

Figure (b)

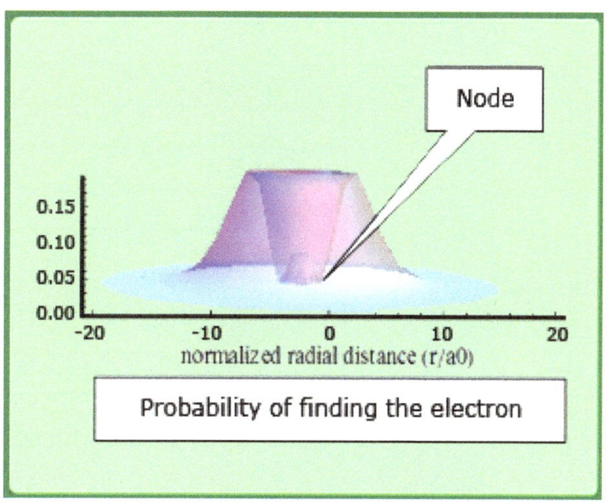

Figure (c)

We have evaluated the wavefunction of the electron in Hydrogen atom as

$$\Psi_{n,\ell,m_l} = \sqrt{\left(\frac{2}{na_0}\right)^3 \frac{(n-\ell-1)!}{2n\{(n+\ell)!\}^3}} e^{-r/na_0} \left(\frac{2r}{na_0}\right)^{\ell} L_{n-\ell-1}^{2\ell+1}\left(\frac{2r}{na_0}\right)$$

$$\times \sqrt{\frac{(2\ell+1)(\ell-m_\ell)!}{4\pi \quad (\ell+m_\ell)!}} p_\ell^{m\ell}(\cos\theta)$$

$$\Rightarrow \Psi_{n,l,m_l}(r,\theta,\phi) = R_{n,l}(r) Y_l^{ml}(\theta,\phi)$$

Where, n => principal quantum number = 1, 2, 3, 4, ...

l => azimuthal quantum number = 0, 1, 2, ..., n-1

and => magnetic quantum number = -l, -l+1, ..., -2, -1, 0, 1, 2, ..., l+1 (=2l+1 values)

And the Energy

$$E_n = -\frac{Z^2\mu e^4}{(4\pi\varepsilon_0)^2 2\hbar^2} \frac{1}{n^2} = \frac{-13.6eV}{n^2} = -\frac{R}{n^2}$$

The energy depends only on n, but the functional form of R(r) depends on both n and l, and so we write the radial wavefunction as $R_{n,l}(r)$ and angular part Y depends on l and m_l.

Principal quantum no.	Orbital quantum no.	Nomenclature	Magnetic quantum no.	Energy
$n = 1$ (*K* shell)	l = 0 (s subshell)	*1s state*	*ml* = 0	-13.6 *eV*
$n = 2$ (*L* shell)	l = 0 (s subshell)	*2s state*	*ml* = 0	-3.4 *eV*
	l = 1 (p subshell)	*2p state*	*ml* = -1	
		2p state	*ml* = 0	
		2p state	*ml* = 1	
$n = 3$ (*M* shell)	l = 0 (s subshell)	*3s state*	*ml* = 0	-1.51 *eV*
	l = 1 (p subshell)	*3p state*	*ml* = -1	
		3p state	*ml* = 0	
		3p state	*ml* = 1	
	l = 2 (d subshell)	*3d state*	*ml* = -2	
		3d state	*ml* = -1	
		3d state	*ml* = 0	
		3d state	*ml* = 1	
		3d state	*ml* = 2	

In the following Figure, we plot the radial functions $R_{n,l}(r)$ the reduced radial functions $U_{n,l}(r) = rR_{n,l}(r)$,and the radial probability densities $p_{n,l}^{rad}(r) = (U_{n,l}(r))^2$ for different states of hydrogen atom.

The first orbit n = 1, l = 0 and m_l = 0, *the K shell*

- In this plot we can see the effect of the factor r in $U_{ls}(r) = rR_{ls}(r)$. Here, $R_{ls}(r)$ goes to $2a_0$ (Bohr radius) at r = 0, $U_{ls}(r)$ goes to zero. The factor of r in $U_{ls}(r)$ also introduces a peak at 1 a_0
- There is no node for 1s state.
- The probability density $p_{ls}^{rad}(r)$ goes to zero at $r = 0$, it means that the probability of finding the electron at the nucleus is zero. $p_{ls}^{rad}(r)$ shows also a peak at $1a_0$ which is the same as Bohr radius and the peak value is 0.54.
- Conceptually, peaks in $p_{n,l}^{rad}(r)$ are radii near which an electron in a state *n, l, m_l* is most likely to be found.
- Because of the gradual decay of $p_{n,l}^{rad}(r)$ on the large-r side of this peak, the calculated mean radial position turns out to be larger than the location of the peak.

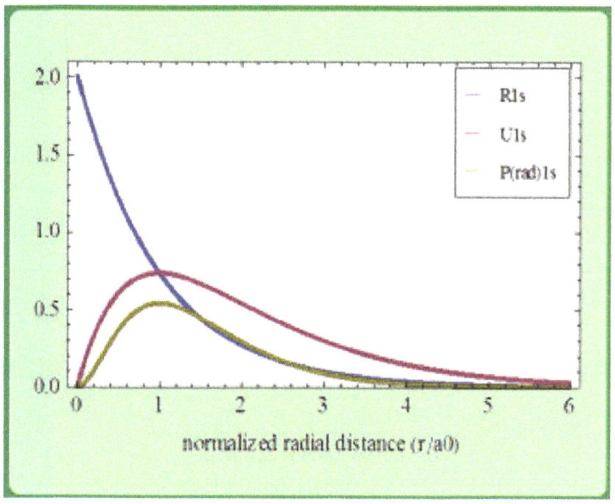

Figure (d)

Following figure is the 1s orbital:

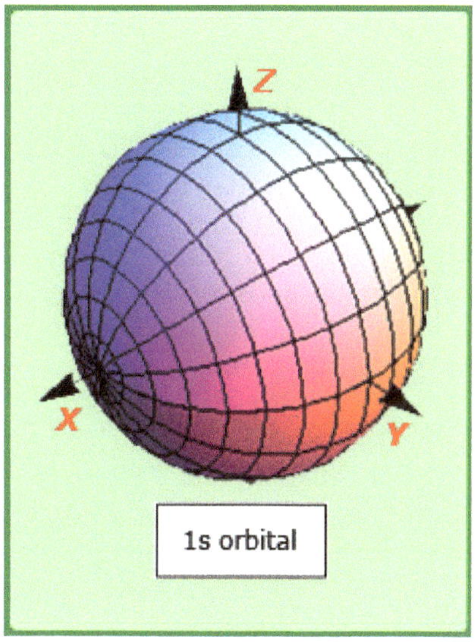

Figure (e)

The second orbit n = 2 (L shell), l = 0, m_l = 0 and l = 1, m_l = 0, ±1

- The 2s radial function in the Figure decays as $r \to \infty$ more gradually than $U_{ls}(r)$. Moreover, the 2s radial probability density has two peaks: one at $0.76a_0$ and next one at $5.24a_0$.
- The height of the peak nearest the nucleus, 0.052, is much smaller than that of the outer peak, 0.19. The presence of a small peak in $p_{2s}^{rad}(r)$ relatively near the origin is called penetration.

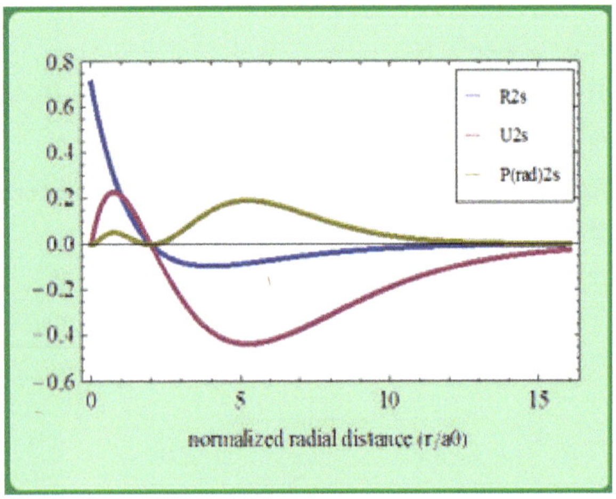

Figure (f)

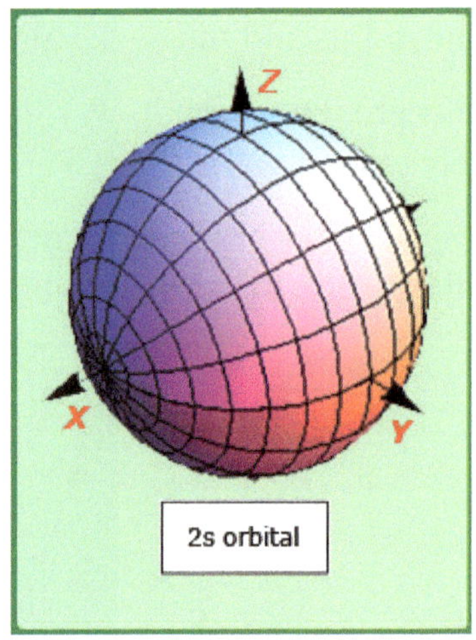

Figure (g)

The 2p functions in this Figure - h differ strikingly near the origin from the 1s and 2s functions. Because of the l = 1 barrier term in the r radial equation, $p_{2p}^{rad}(r)$ rises from zero.

- $U_{2p}(r)$ has a classically forbidden region near the origin.

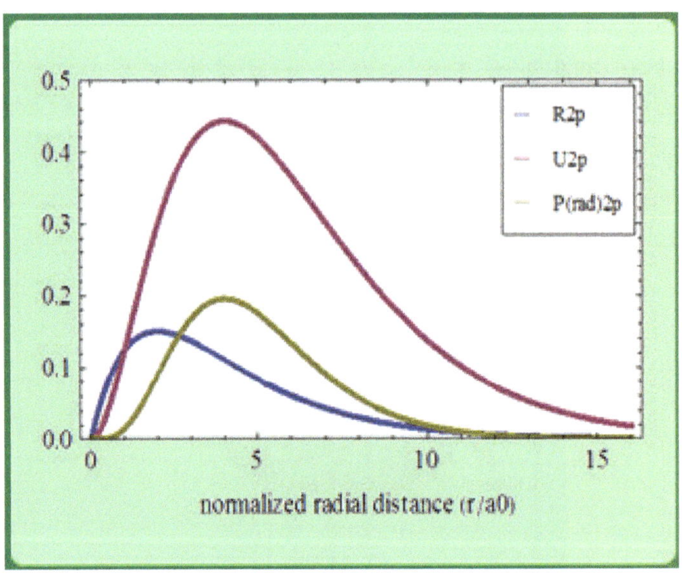

Figure (h)

Following figures are three $p_x, p_y and p_z$ orbitals:

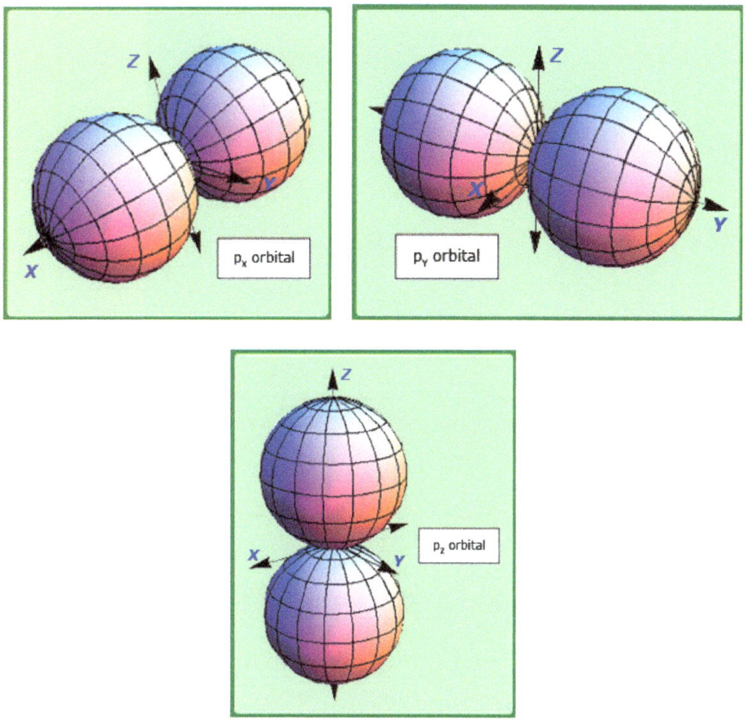

Figure (i)

To compare the differences between the functions $U_{n,l}(r) = rR_{n,l}(r)$ and $p_{n,l}^{rad}(r) = (U_{n,l}(r))^2$ for the n = 1 and n = 2 states of hydrogen, they are plotted on the same graph as shown in the following Figure - j and Figure - k:

- The increase in l by 1 from 2s to 2p states means that neither $U_{2p}(r)$ nor $p_{2p}^{rad}(r)$ have nodes. Asymptotically, both functions decay at the same rate as their 2s counterparts, because the decay constant for a pure- Coulomb potential energy doesn't depend on l.
- The sole peak in $p_{2p}^{rad}(r)$ occurs at r = $4a_o$, where its value is 0.19. Interestingly, the mean radius of 2p state = $5a_o$ is $1a_o$ smaller than that of the 2s state.
- In the plots of reduced radial functions $U_{1s}(r)$, $U_{2p}(r)$,
note the differences in these functions rise from 0. The functions $U_{1s}(r)$ and $U_{2s}(r)$ rise like r, in which $U1s(r)$ is faster than $U_{2s}(r)$ while $U_{2s}(r)$ rises like r^2.

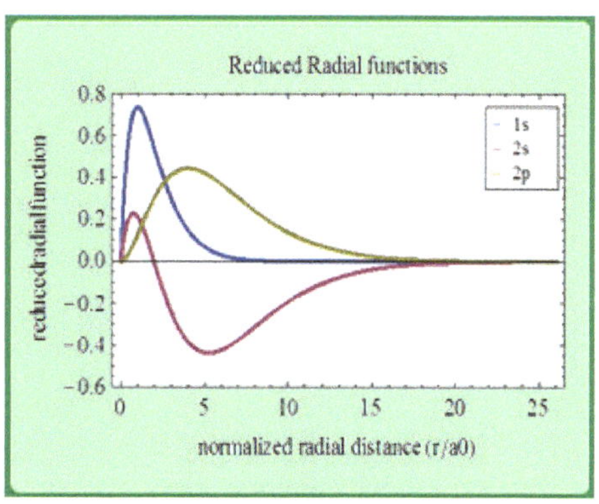

Figure (j)

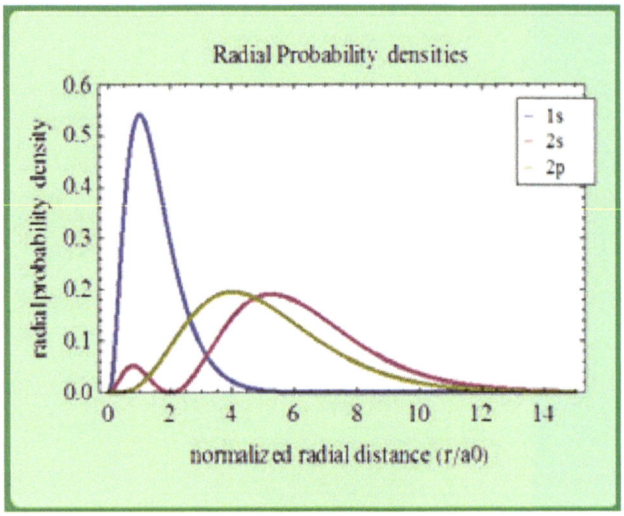

Figure (k)

The third orbit n = 3 (M shell), l = 0, m_l = 0 and l = 1, m_l = 0, ±1 and l = 2, m_l = 0, ±1, ±2

- The general features are from the following Figure - l:

  $U_{3s}(r)$ has two nodes, $U_{3p}(r)$ has one, and $U_{3d}(r)$ has none.

- The 3*s* radial probability density has two regions fairly near the origin where there is a small but nonzero chance of finding the electron.

- The 3*s* state is the most penetrating, because its radial function is unaffected by a barrier, while the 3d, which suffers the strongest barrier, is the least penetrating.

- Since the number of nodes = *n* - *l* – 1 decreaseswith increasing *l*, small *l* radial functions have the most nodes.

- Since these nodes are radii where the probability density is zero, radial probability densities for low- *l* are characterized by small regions of non-zero probability very near the nucleus.

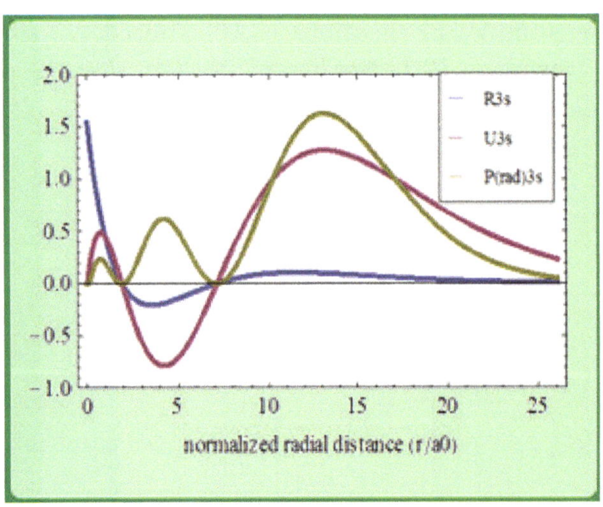

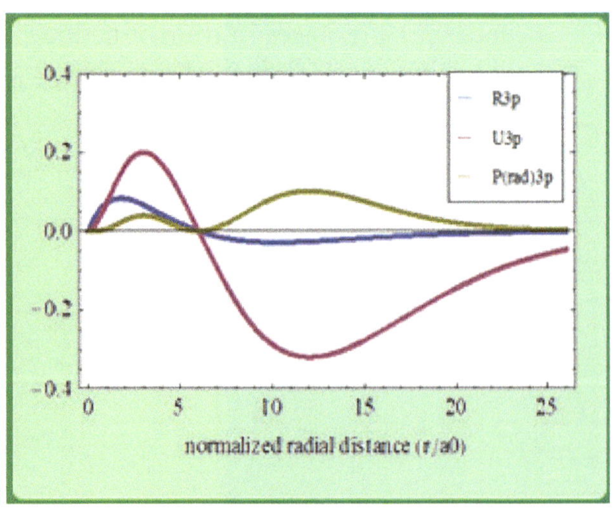

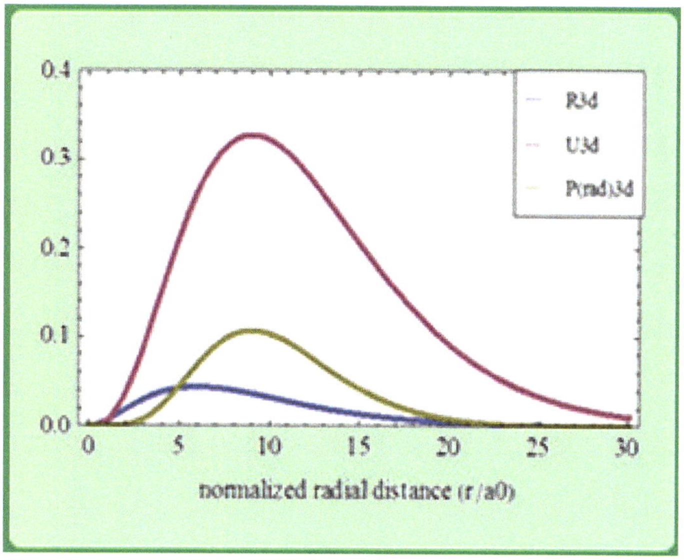

Figure (l)

The following Figure - m show the comparison of reduced radial functions and probability densities of n = 3 states:

- The n = 3 states show that as l increases, the outermost peak moves inward, to smaller radial distance.
- The magnitude of each peak increases as the radial distance increases. This increase occurs because as l increases by one unit, the probability density, which must remain normalized to unity, looses one inner peak.
- In spite of all these variations, the energies of all these n = 3 states is the same, E_3 = 1.51 eV.
- The range of r values required to plot all of the radial probability density for the n = 3 states is more than twice for states with n < 3.

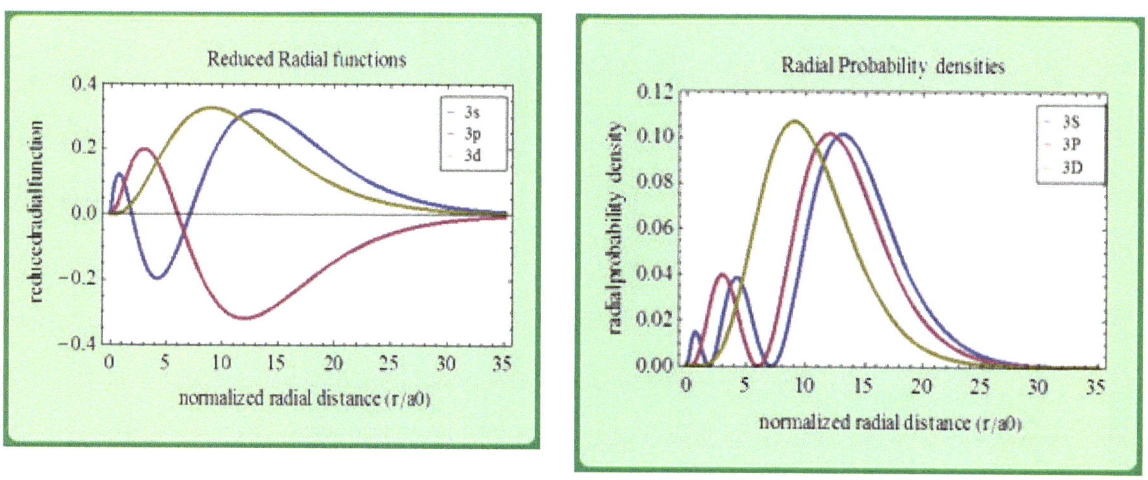

Figure (m)

The five d - orbitals are shown in Figure - n:

Figure (n)

Following Figure - o show the comparison of the probability densities for n = 1, 2, 3, 4 states.

- All s-state radial probability densities for n = 1, 2, 3, and 4 behave the same way near the origin. None suffer the l barrier, so all have peaks near the origin.

- Because the number of nodes in an s state function increases by one with each unit increase in n, the number of peaks also increases by one.
- Only for the 1s state has a large peak near the origin; the magnitudes of the near-origin peaks in radial probability density for n > 1 decrease rapidly with increasing n.
- Consequently the probability of finding the electron far from the nucleus increases appreciable as n increases.

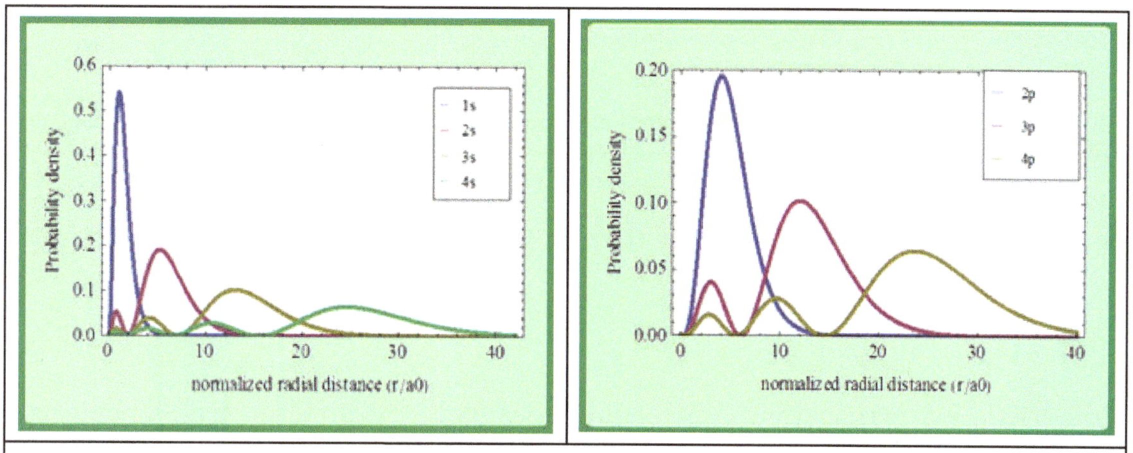

Figure (o)

Main features to remember

- l = 0, s-states do not vanish at r = 0.
- $l \neq 0$, states vanish at r = 0 and have their maximum probability amplitude further out with increasing l.
- The size, position of peak probability, scales with $\sim n^2$.
- The l = 0 function crosses the axis (n – 1) times i.e. it has (n – 1) nodes.
- l = 1 has (n – 2) nodes and so on.
- Maximum l = n – 1 has no nodes (except at r = 0)

Stern–Gerlach Experiment

The Stern–Gerlach experiment demonstrated that the spatial orientation of angular momentum is quantized. It demonstrated that atomic-scale systems have intrinsically quantum properties, and that measurement in quantum mechanics affects the system being measured. In the original experiment, silver atoms were sent through a non-uniform magnetic field, which deflected them before they struck a detector screen. Other kinds of particles can be used. If the particles have a magnetic moment related to their spin angular momentum, the magnetic field gradient deflects them from a straight path. The screen reveals discrete points of accumulation rather than a continuous distribution, owing to the quantum nature of spin. Historically, this experiment was decisive in convincing physicists of the reality of angular momentum quantization in all atomic-scale systems.

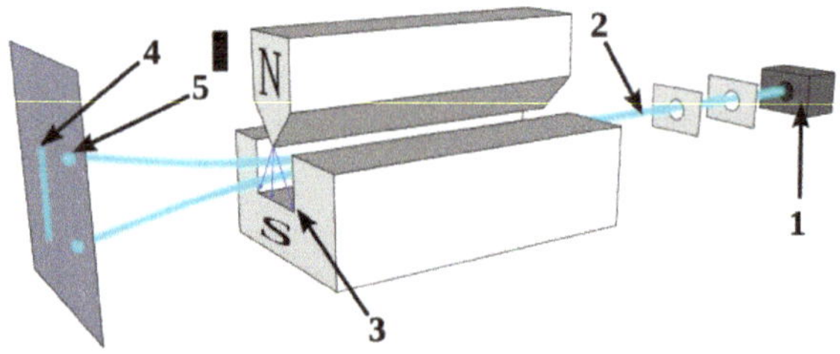

Stern–Gerlach experiment: silver atoms travel through an inhomogeneous magnetic field and are deflected up or down depending on their spin. 1: furnace. 2: beam of silver atoms. 3: inhomogeneous magnetic field. 4: expected result . 5: what was actually observed

The experiment was first conducted by the German physicists Otto Stern and Walther Gerlach, in 1922.

Basic Theory and Description

The Stern–Gerlach experiment involves sending a beam of particles through an inhomogeneous magnetic field and observing their deflection. The results show that particles possess an intrinsic angular momentum that is closely analogous to the angular momentum of a classically spinning object, but that takes only certain quantized values. Another important result is that only one component of a particle's spin can be measured at one time, meaning that the measurement of the spin along the z-axis destroys information about a particle's spin along the x and y axis.

The experiment is normally conducted using electrically neutral particles or atoms. This avoids the large deflection to the orbit of a charged particle moving through a magnetic field and allows spin-dependent effects to dominate. For example, observation of the Stern–Gerlach effect with free electrons is infeasible. If the particle is treated as a classical spinning dipole, it will precess in a magnetic field because of the torque that the magnetic field exerts on the dipole. If it moves through a homogeneous magnetic field, the forces exerted on opposite ends of the dipole cancel each other out and the trajectory of the particle is unaffected. However, if the magnetic field is inhomogeneous then the force on one end of the dipole will be slightly greater than the opposing force on the other end, so that there is a net force which deflects the particle's trajectory. If the particles were classical spinning objects, one would expect the distribution of their spin angular momentum vectors to be random and continuous. Each particle would be deflected by a different amount, producing some density distribution on the detector screen. Instead, the particles passing through the Stern–Gerlach apparatus are deflected either up or down by a specific amount. This was a measurement of the quantum observable now known as spin angular momentum, which demonstrated possible outcomes of a measurement where the observable has a discrete set of values or point spectrum. Although some discrete quantum phenomena, such as atomic spectra, were observed much earlier, the Stern–Gerlach experiment allowed scientists to observe separation between discrete quantum states for the first time in the history of science.

By now it is known theoretically that quantum angular momentum *of any kind* has a discrete spectrum, which is sometimes imprecisely expressed as "angular momentum is quantized".

If the experiment is conducted using charged particles like electrons, there will be a Lorentz force that tends to bend the trajectory in a circle. This force can be cancelled by an electric field of appropriate magnitude oriented transverse to the charged particle's path.

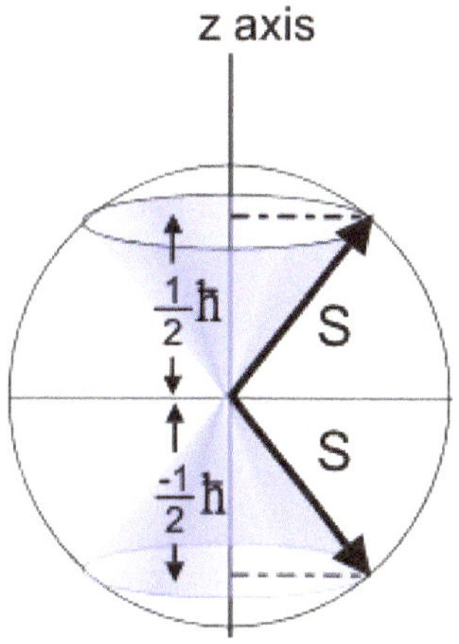

Spin values for fermions

Electrons are spin-½ particles. These have only two possible spin angular momentum values measured along any axis, $+\hbar/2$ or $-\hbar/2$, a purely quantum mechanical phenomenon. Because its value is always the same, it is regarded as an intrinsic property of electrons, and is sometimes known as "intrinsic angular momentum" (to distinguish it from orbital angular momentum, which can vary and depends on the presence of other particles).

To describe the experiment with spin +½ particles mathematically, it is easiest to use Dirac's bra–ket notation. As the particles pass through the Stern–Gerlach device, they are being observed by the detector which resolves to either spin up or spin down. These are described by the angular momentum quantum number j, which can take on one of the two possible allowed values, either $+\hbar/2$ or $-\hbar/2$. The act of observing (measuring) the momentum along the z axis corresponds to the operator J_z. In mathematical terms, the initial state of the particles is

$$|\psi\rangle = c_1 \left| \psi_{j=+\frac{\hbar}{2}} \right\rangle + c_2 \left| \psi_{j=-\frac{\hbar}{2}} \right\rangle,$$

where constants c_1 and c_2 are complex numbers. This initial state spin can point in any direction. The squares of the absolute values ($|c_1|^2$ and $|c_2|^2$) determine the probabilities that for a system in the initial state $|\psi\rangle$ one of the two possible values of j is found after the measurement is made. The constants must also be normalized in order that the probability of finding either one of the values be unity. However, this information is not sufficient to determine the values of c_1 and c_2, because they are complex numbers. Therefore, the measurement yields only the squared magnitudes of the constants, which are interpreted as probabilities.

Sequential Experiments

If we link multiple Stern–Gerlach apparatuses, we can clearly see that they do not act as simple selectors, but alter the states observed (as in light polarization), according to quantum mechanical law:

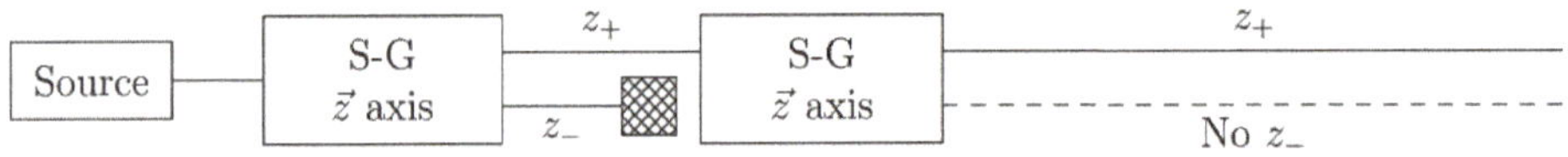

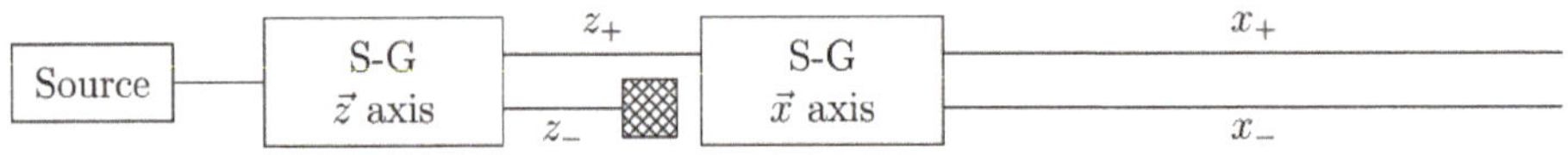

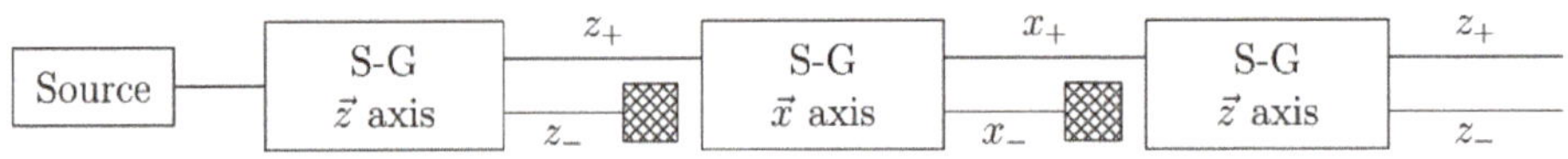

History

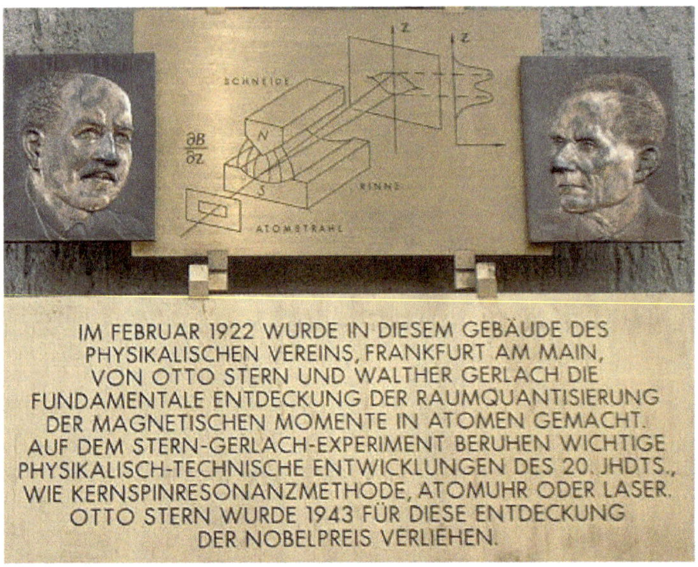

A plaque at the Frankfurt institute commemorating the experiment

The Stern–Gerlach experiment was performed in Frankfurt, Germany in 1922 by Otto Stern and Walther Gerlach. At the time, Stern was an assistant to Max Born at the University of Frankfurt's Institute for Theoretical Physics, and Gerlach was an assistant at the same university's Institute for Experimental Physics.

At the time of the experiment, the most prevalent model for describing the atom was the Bohr model, which described electrons as going around the positively charged nucleus only in certain discrete atomic orbitals or energy levels. Since the electron was quantized to be only in certain positions in space, the separation into distinct orbits was referred to as space quantization. The Stern–Gerlach experiment was meant to test the Bohr–Sommerfeld hypothesis that the direction of the angular momentum of a silver atom is quantized.

Note that the experiment was performed several years before Uhlenbeck and Goudsmit formulated their hypothesis of the existence of the electron spin. Even though the result of the Stern–Gerlach experiment has later turned out to be in agreement with the predictions of quantum mechanics for a spin-$1/2$ particle, the experiment should be seen as a corroboration of the Bohr–Sommerfeld theory.

In 1927, T.E. Phipps and J.B. Taylor reproduced the effect using hydrogen atoms in their ground state, thereby eliminating any doubts that may have been caused by the use of silver atoms. (In 1926 the non-relativistic Schrödinger equation had incorrectly predicted the magnetic moment of hydrogen to be zero in its ground state. To correct this problem Wolfgang Pauli introduced "by hand", so to speak, the 3 Pauli matrices which now bear his name, but which were later shown by Paul Dirac in 1928 to be intrinsic in his relativistic equation.)

The experiment was first performed with an electromagnet that allowed the non-uniform magnetic field to be turned on gradually from a null value. When the field was null, the silver atoms were deposited as a single band on the detecting glass slide. When the field was made stronger, the middle of the band began to widen and eventually to split into two, so that the glass-slide image looked like a lip-print, with an opening in the middle, and closure at either end. In the middle, where the magnetic field was strong enough to split the beam into two, statistically half of the silver atoms had been deflected by the non-uniformity of the field.

Importance

The Stern–Gerlach experiment strongly influenced later developments in modern physics:

- In the decade that followed, scientists showed using similar techniques, that the nuclei of some atoms also have quantized angular momentum. It is the interaction of this nuclear angular momentum with the spin of the electron that is responsible for the hyperfine structure of the spectroscopic lines.
- In the 1930s, using an extended version of the Stern–Gerlach apparatus, Isidor Rabi and colleagues showed that by using a varying magnetic field, one can force the magnetic moment to go from one state to the other. The series of experiments culminated in 1937 when they discovered that state transitions could be induced using time varying fields or RF fields. The so-called Rabi oscillation is the working mechanism for the Magnetic Resonance Imaging equipment found in hospitals.
- Norman F. Ramsey later modified the Rabi apparatus to increase the interaction time with the field. The extreme sensitivity due to the frequency of the radiation makes this very useful for keeping accurate time, and it is still used today in atomic clocks.
- In the early sixties, Ramsey and Daniel Kleppner used a Stern–Gerlach system to produce a beam of polarized hydrogen as the source of energy for the hydrogen Maser, which is still one of the most popular atomic clocks.
- The direct observation of the spin is the most direct evidence of quantization in quantum mechanics.
- The Stern–Gerlach experiment has become a paradigm of quantum measurement.

In 1922, Stern and Gerlach launched a beam of Silver (Ag) atoms through a non-uniform magnetic field and detected them at a screen as shown in the figure below.

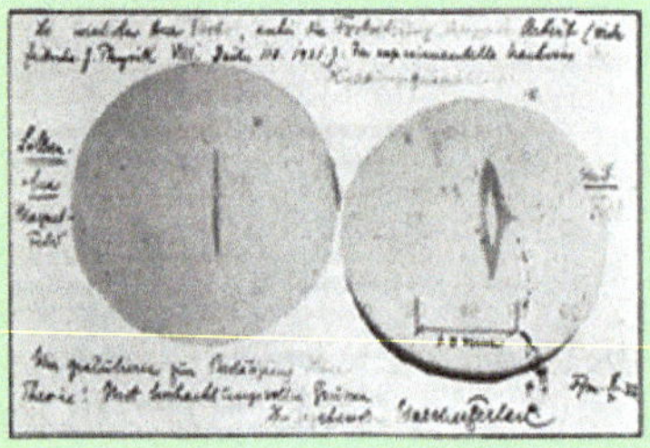

Gerlach's postcard, dated Feb, 1922, to Niels Bohr showing a photograph of the splitting.

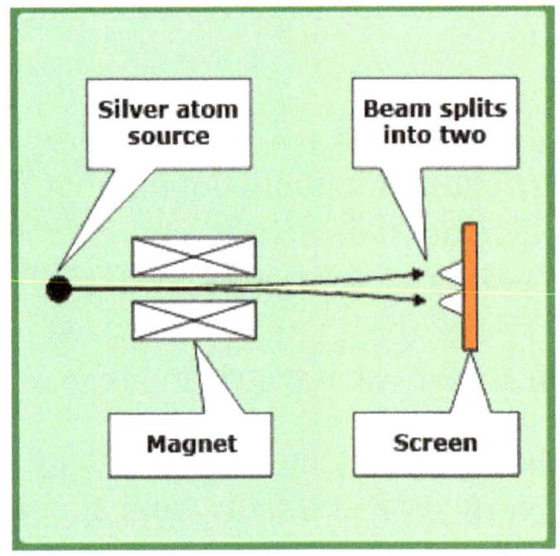

- Classical physics does not predict distinct spots. Distinct spots mean the magnetic moment is quantized.
- This can not be due to orbital magnetic moment because orbital magnetic quantum number is odd, we would expect odd number of spots.
- This marked the discovery of the electron spin. A new type of angular momentum, with a quantum number that can take on only two values:

$$s = \tfrac{1}{2}, m_s = \pm \tfrac{1}{2}$$

This new kind of angular momentum is known as SPIN.

This spin sometimes visualize by the classical spin picture as shown in the following figure.

However, this "spinning" ball picture shown in Figure is not realistic, because it would require that the tiny electron be spinning so fast that parts would travel faster than c. So the spin is another type of angular momentum.

So, we need FOUR quantum numbers to specify the electronic state of a hydrogen atom

n, l, m_l, m_s (where $m_s = -\tfrac{1}{2}$ and $m_s = +\tfrac{1}{2}$).

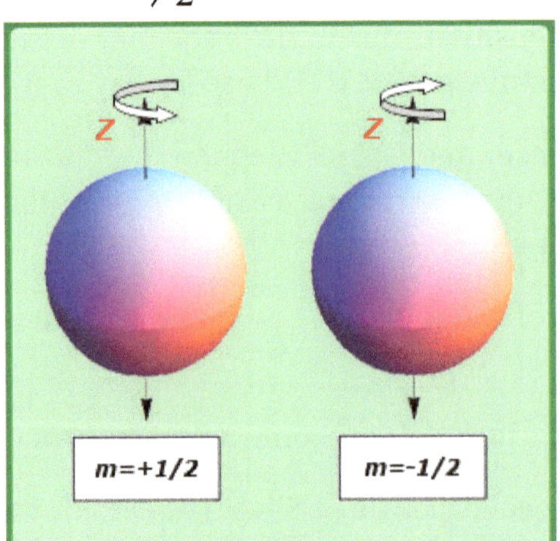

However, this "spinning" ball picture shown in the figure below is not realistic, because it would require that the tiny electron be spinning so fast that parts would travel faster than c. So the spin is another type of angular momentum.

So, we need FOUR quantum numbers to specify the electronic state of a hydrogen atom n, l, m_l, m_s (where $m_s = -\frac{1}{2}$ and $m_s = +\frac{1}{2}$).

References

- Phipps, T.E.; Taylor, J.B. (1927). "The Magnetic Moment of the Hydrogen Atom". Physical Review. 29 (2): 309–320. Bibcode:1927PhRv...29..309P. doi:10.1103/PhysRev.29.309
- Gerlach, W.; Stern, O. (1922). "Der experimentelle Nachweis der Richtungsquantelung im Magnetfeld". Zeitschrift für Physik. 9: 349–352. Bibcode:1922ZPhy....9..349G. doi:10.1007/BF01326983
- Gerlach, W.; Stern, O. (1922). "Der experimentelle Nachweis des magnetischen Moments des Silberatoms". Zeitschrift für Physik. 8: 110–111. Bibcode:1922ZPhy....9..349G. doi:10.1007/BF01329580
- Stern, O. (1921). "Ein Weg zur experimentellen Pruefung der Richtungsquantelung im Magnetfeld". Zeitschrift für Physik. 7: 249–253. Bibcode:1921ZPhy....7..249S. doi:10.1007/BF01332793
- French, A.P., Taylor, E.F. (1979). An Introduction to Quantum Physics, Van Nostrand Reinhold, London, ISBN 0-442-30770-5, pp. 428–442
- Weinert, F. (1995). "Wrong theory—right experiment: The significance of the Stern–Gerlach experiments". Studies in History and Philosophy of Modern Physics. 26B: 75–86. doi:10.1016/1355-2198(95)00002-B

4

Fundamentals of Molecule Formation

Chemical bonding, ionic bonding and other processes can be better understood with the help of quantum mechanics. Molecular formation and Coulomb force are one of the significant and important topics related to molecular physics. The following chapter unfolds its crucial aspects in a critical yet systematic manner.

Quantum Theory of Molecule Formation

It only became possible with the aid of quantum mechanics to attain a fundamental understanding for the formation of molecule.

Even in the case of ionic bonding, basic new insights have been obtained through quantum theory.

First, for example, it must be understood why the ions form in the first place, and why the electron which is transferred from sodium to chlorine thus finds an energetically more favorable state.

In the following, we will develop some important basic ideas for the quantum theory of chemical bonding. However, Physics & Chemistry are still far away from a complete solution to these problems.

To understand chemical bonding, the interactions of several particles must be taken into account: given n atomic nuclei and m electrons, one would have to find the complete wavefunction and the corresponding energies of the total system.

It is useful to keep in mind that the nuclear masses are much greater than those of the electrons. Thus the electronic motions are much faster than nucleus.

Then we may ignore the motion of the nuclei and treat them as fixed.

In atomic physics, we were able to obtain much information from spectroscopic observations and could direct our attention to both the ground states and the excited states.

In the study of chemical bonding, the determination of the wavefunction of the ground state of the particular molecule plays a more important role.

The Hydrogen – Molecule Ion H_2^+

Certainly the simplest case of chemical bonding occurs in the hydrogen molecule ion H_2^+. This species is observed as a bound state in gas discharges in a hydrogen atmosphere, in such a gas discharge, the hydrogen molecule loses one electron.

The bonding energy, equivalent to the dissociation energy, has been determined to be 2.65 eV.

We are dealing with two nuclei and one electron. If the nuclei are far removed from one another, we can imagine that the electron is localized on one nucleus or other.

The wave functions are those of the hydrogen atomic ground state.

So the Hamiltonian for nucleus a,

$$\underbrace{\left(-\frac{\hbar^2}{2m}\nabla^2-\frac{e^2}{r_a}\right)}_{H_\sigma}\phi_a(r_a)=E_a^0\phi_a(r_a)$$

And correspondingly for nucleus b,

$$H_b\phi_b(r_b)=E_b^0\phi_b(r_b)$$

So, $E_a^0=E_b^0=E^0$

If we let the nuclei approach one another, the electron, which was, for example, at first attached to nucleus a, will experience the attractive Coulomb force of nucleus b.

Conversely, an electron which was at first bound to nucleus b will experience the attractive Coulomb force of nucleus a. We must therefore set up a Schrodinger Equation which contains the Coulomb potential for both.

Further in order to calculate the total energy of the system, we must take into account the Coulomb repulsion between nuclei. The additional energy

$\left(e^2/R_{ab}\right)$ is not directly related to the energy of the electron, it will only produce a constant shift of all the energy eigen values. We will introduce it at the end.

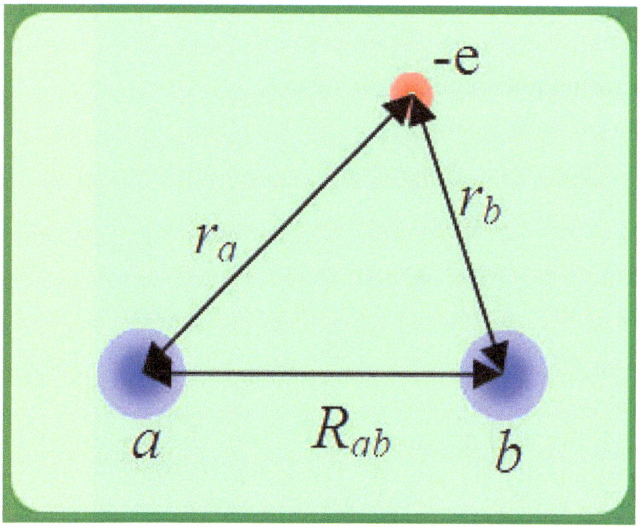

Referring to the figure above, let us define,

$$R_{ab}=|R_a-R_b|\qquad r_a=|r-R_a|\qquad r_b=|r-R_b|$$

Thus, Schrödinger equation, then

$$\left(-\frac{\hbar^2}{2m}\nabla^2-\frac{e^2}{r_a}-\frac{e^2}{r_b}\right)\psi=E\psi$$

Here E and ψ are energy eigen value and the wavefunction for the whole system respectively and are yet to be calculated.

We will use the perturbation method. In principle, the electron could be in the neighborhood of nucleus a or of nucleus b, with the same energy. These two states ϕ_a and ϕ_δ are thus degenerate. Now however, the other nucleus, from which the electron is by chance more distant, acts, as a perturbation to the electronic state. We thus expect that the degeneracy will be lifted by this perturbation.

In the presence of the degeneracy, we take the total wavefunction is the linear combination of the hydrogen wavefunctions defined in equation above and defined as,

$$\psi=c_1\phi_a+c_2\phi_b$$

Where *c1* and *c2* are the coefficients. This is generally known as linear combination of the atomic orbital (LCAO) and ψ is known as the molecular orbital.

So, putting this, in equation above, and rearranging we get,

$$\left(H_a-\frac{e^2}{r_b}\right)c_1\phi_a+\left(H_b-\frac{e^2}{r_b}\right)c_2\phi_b=E(c_1\phi_a+c_2\phi_b)$$

Where

$$H_a=-\frac{\hbar^2}{2m}\nabla^2-\frac{e^2}{r_a} \text{ and } H_b=-\frac{\hbar^2}{2m}\nabla^2-\frac{e^2}{r_b}$$

Substituting the values from equations above, we get,

$$\left(E^0-\frac{e^2}{r_b}\right)c_1\phi_a+\left(E^0-\frac{e^2}{r_a}\right)c_2\phi_b=Ec_1\phi_a+Ec_2\phi_b$$

$$\Rightarrow\left(\underbrace{E^0-E}_{-\Lambda E}-\frac{e^2}{r_b}\right)c_1\phi_a+\left(\underbrace{E^0-E}_{-\Lambda E}-\frac{e^2}{r_a}\right)c_2\phi_b=0$$

While ϕ_a and ϕ_b are functions of position, the coefficients are independent of positions.

We assume that the functions ϕ_a and ϕ_b are real, as in the case of hydrogen atom ground state wavefunctions, and the functions ϕ_a and ϕ_b are not orthogonal since the electron is associated with two different nuclei *a* and *b*. So,

$$\int\phi_a\phi_b dV=s \quad \text{and} \quad \int\phi_a\phi_b dV=\int\phi_a\phi_b dV=1$$

Now, Equation 6 is multiplied by ϕ_a

$$\int \phi_a \left(-\Delta E - \frac{e^2}{r_b} \right) c_1 \phi_a dV + c_2 \int \phi_a \left(-\Delta E - \frac{e^2}{r_a} \right) \phi_b dV = 0$$

$$\Rightarrow -\Delta E c_1 - c_1 \int \phi_a(r_a) \frac{e^2}{r_b} \phi_a(r_a) dV + c_2(-\Delta E) \int \phi_a \phi_b dV - c_2 \int \phi_a(r_a) \frac{e^2}{r_2} \phi_a(r_b)\, dV = 0$$

$$\Rightarrow (-\Delta E c_1 - c_1 C) + (-\Delta E c_2 S - D c_2) = 0$$

$$\Rightarrow (-\Delta E - C) c_1 + (-\Delta E.S - D) c_2 = 0$$

Where,

$$+\int \phi_a(r_a) \frac{e^2}{r_b} \phi_a(r_a) dV = c$$

$$+\int \phi_a(r_a) \frac{e^2}{r_b} \phi_b(r_b) dV = D$$

Equation above is multiplied by ϕ_b and following the same procedure we get,

$$(-\Delta E.S - D) c_1 + (\Delta E - C) c_2 = 0$$

So, to get the solution, following determinant from equations 8 and 9 should vanish.

The determinant:

$$\begin{pmatrix} \Delta E + C & \Delta E.S + D \\ \Delta E.S + D & \Delta E + C \end{pmatrix} \begin{pmatrix} c_1 \\ c_2 \end{pmatrix} = 0$$

So,

$$(\Delta E + C^2) - (\Delta E.S + D)^2 = 0$$

$$\Rightarrow \Delta E^2 + C^2 + 2\Delta E C - \Delta E^2 S^2 - D^2 - 2\Delta E D S = 0$$

$$\Rightarrow \Delta E^2 (1 - S) + 2(C - DS)\Delta E + (c^2 - D^2) = 0$$

$$\Rightarrow \Delta E = \frac{-2(C - DS) \pm \sqrt{4(C - DS)^2 - 4(1 - S^2)(C^2 - D^2)}}{2(1 - s^2)}$$

$$= \frac{-(C - DS) \pm \sqrt{C^2 + D^2 S^2 - 2CDS - C^2 + D^2 + C^2 S^2 - D^2 S^2}}{(1 - S^2)}$$

$$= \frac{-(C - DS) \pm \sqrt{(D - CS)^2}}{(1 - S^2)}$$

$$= \frac{-(C - DS) \pm (D - CS)}{(1 - S^2)}$$

Two solutions

(i) $\dfrac{-C+DS-D+CS}{1-S^2} = -\dfrac{C+D}{1+S}$

(ii) $\dfrac{-C+DS+D-CS}{1-S^2} = -\dfrac{C-D}{1-S}$

So,

$$\Delta E = -\frac{c \pm D}{1 \pm s}$$

If we take, solution in equation above and then,

$$Substituting \left[\Delta E = -\frac{C+D}{1+S} \right]$$

$$(\Delta E + C)c_1 + (\Delta E.S + D)c_2 = 0$$

$$\Rightarrow \left(-\frac{C+D}{1+S} + C \right) c_1 + \left(-\frac{(C+D)S}{1+S} + D \right) c_2 = 0$$

$$\Rightarrow \left(\frac{-C-D+C+CS}{1+S} \right) c_1 + \left(\frac{-CS-DS+C+DS}{1+S} \right) c_2 = 0$$

$$\Rightarrow \left(\frac{CS-D}{1+S} \right) c_1 + \left(\frac{D-CS}{1+S} \right) c_2 = 0$$

$$\Rightarrow \left(\frac{CS-D}{1+S} \right) c_1 - \left(\frac{CS-D}{1+S} \right) c_2 = 0$$

So, $c_1 = c_2 = c$ and thus, $\psi_+ = c(\phi_a + \phi_b)$

$$E^0 - E = -\Delta E$$

$$\Rightarrow E = E^0 - \frac{C+D}{1+S} = E.$$

Now if we take, solution in equation above and

$$Substituting \left[\Delta E = -\frac{C-D}{1-S} \right]$$

$$(\Delta E.S + D)c_1 + (\Delta E + C)c_2 = 0$$

$$\Rightarrow \left(-\frac{(C-D)S}{1-S} + D \right) c_1 + \left(-\frac{(C-D)}{1-S} + C \right) c_2 = 0$$

$$\Rightarrow \left(\frac{-CS+DS+D-DS}{1-S} \right) c_1 + \left(\frac{-C+D+C-CS}{1-S} \right) c_2 = 0$$

$$\Rightarrow \left(\frac{D-CS}{1-S} \right) c_1 - \left(\frac{D-CS}{1-S} \right) c_2 = 0$$

So, $c_1 = -c_2 = c$ and thus $\psi\_ = c(\phi_a - \phi_b)$

$$E = E^0 - \frac{C-D}{1-S} = E_-$$

Let us understand the meaning of this wavefunctions ψ_+ and ψ_-. Figure below shows the plotting of wavefunctions ϕ_a and ϕ_b as well as $(\phi_a + \phi_b)$.

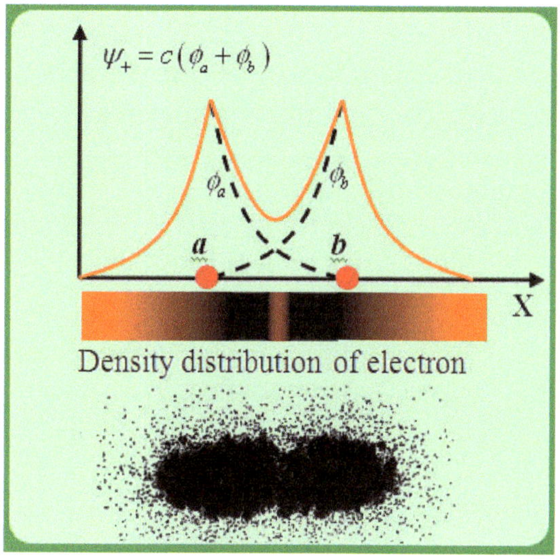

The following observations can be made.

a. Symmetric wavefunction ψ_+ is formed by $\phi_a + \phi_b$. Because of the overlap, the occupation probability for ψ_+ between two nuclei is increased.

b. Thus the density distribution of electron in the ψ_+ state shown in the lower part of figure increases as the distance between a and b decreases.

Antisymmetric wavefunction ψ_- is formed from $\phi_a - \phi_b$. The occupation probability is clearly zero in the plane of symmetry.

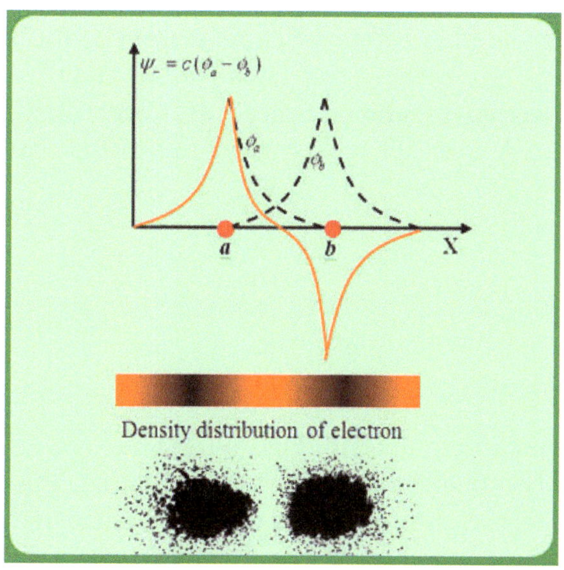

As can be seen from the above figures that the distance between a and b is an important factor to have finite overlap between the ϕ_a and ϕ_b.

Now we will concentrate on the energy values E_+ and E_-. To do that we have to understand the meaning of the quantities S, C and D. We will try to evaluate these quantities through the diagrams.

The quantity $\int \phi_a \phi_b dV = S$

From the figure below it is clear that this quantity S is considerable when Rab is sufficiently close and S increases when Rab decreases. The quantity S is known as overlap integral.

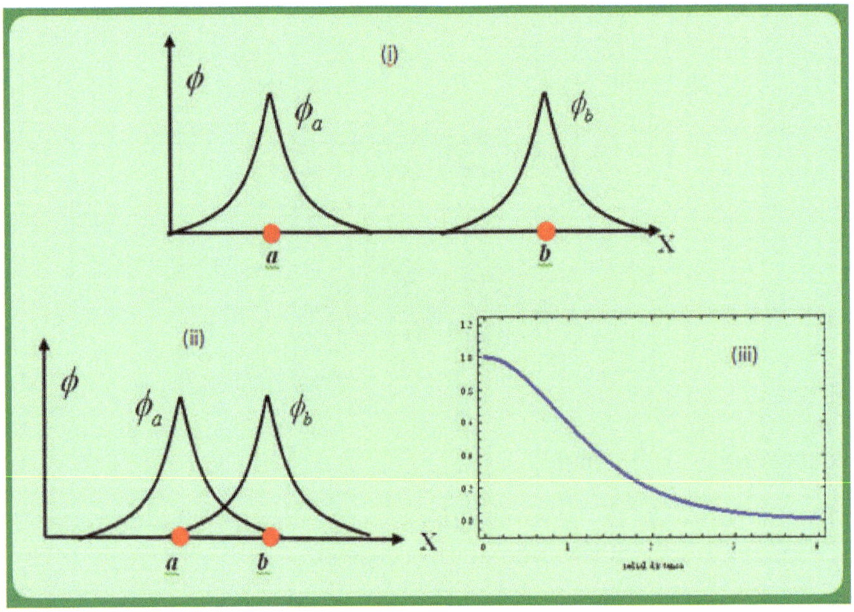

Figure-(i) represents the 1s atomic orbital of hydrogen for two nuclei a and b.

Figure-(ii) shows that when the nuclei are close enough the overlap increases.

Figure-(iii) plots the value of overlap integral S with respect to the internuclear distance Rab. When both the nuclei are coming closer the overlap increases and the value of S increases. When they fall on each other, the overlap integral becomes 1. However, this situation does not arise due to the nuclear-nuclear repulsion term. We will visualize this picture later.

$$C = +\int \phi_a(r_a) \frac{e^2}{r_b} (r_a) dV$$

$$= +\int \phi_b(r_b) \frac{e^2}{r_b} \phi_b(r_b) dV$$

his term is known as the direct coulomb term. Figure-(i) shows the orbital $|\phi_a^2|$ and the coulomb attraction potential. As the two nuclei approaches to each other, the overlap between them increases. Figure-(ii) plots the term C with respect to the internuclear distance *Rab*. Here also we see that, this term is appreciable when the both the nuclei are close enough.

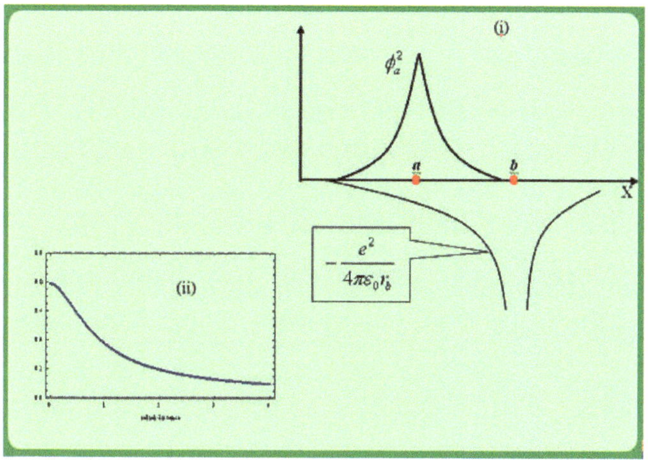

$$D = +\int \phi_a(r_a)\frac{e^2}{r_a}\phi_b(r_a)dV$$

$$= +\int \phi_a(r_a)\frac{e^2}{r_b}\phi_b(r_a)dV$$

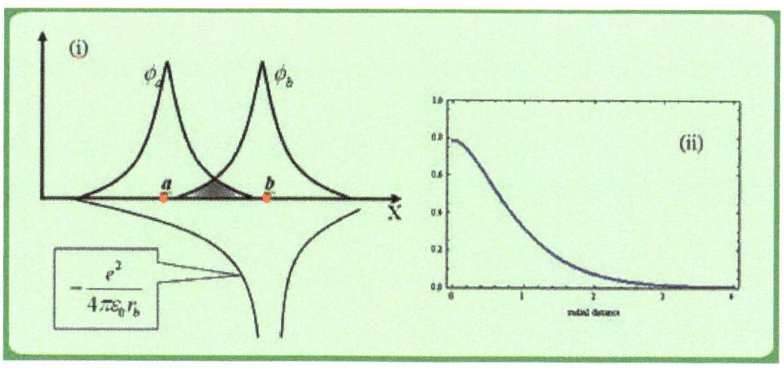

The term D is known as cross integral term. The meaning of this is that there should be overlap between ϕ_a, ϕ_b and coulomb attraction term $(-\frac{e^2}{4\pi\varepsilon_0 r_0})$

as shown in figure (i). The value of this integral is plotted in figure (ii). It is clear from this figures that when there is no overlap between , ϕ_a and ϕ_b , the value of D goes to zero.

In the figure below (i), the quantity S, C, D are plotted together to have a comparison. It can be seen that all these quantities depend on the R_{ab}.

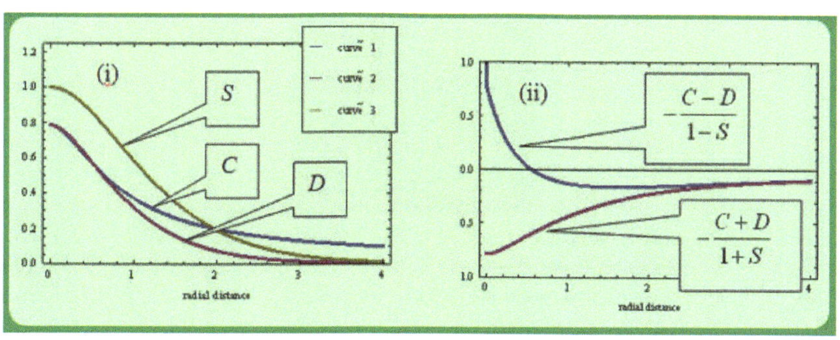

In the above figure (ii), the quantities $-\frac{C+D}{1+S}$ and $-\frac{C-D}{1-S}$ are plotted with respect to the internuclear distance *Rab*. As can be seen, from this plot that these quantities are degenerate when the two nuclei are far away from each other. If two nuclei approach one another the quantities split up in a way depending on whether symmetric or antisymmetric / bonding or antibonding. In Symmetric case energy reduces whereas in Antisymmetric case energy increases.

Now, if we add nuclear energy, the energy diagram is shown in the following figure symmetric case there is a minimum, so it forms a bound state. Antibonding is not a stable state.

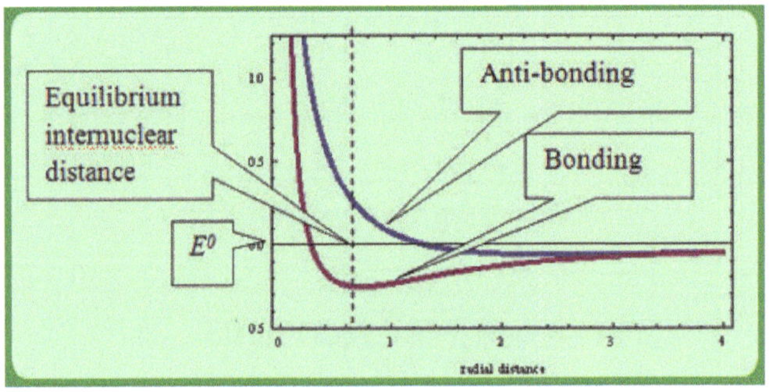

Physical meaning: in the bound state the probability density of electron is quite large in the middle, profit Coulomb attraction from both the nuclei, energy goes down. Whereas in the antibonding case, only electron experiences attractive force of one nucleus at a time. The internuclear distance where the energy is having a minimum is known as equilibrium internuclear distance. Thus, this molecule H2+ - ion is stable with the bonding energy lower than the individual hydrogen atom.

Molecular Orbital

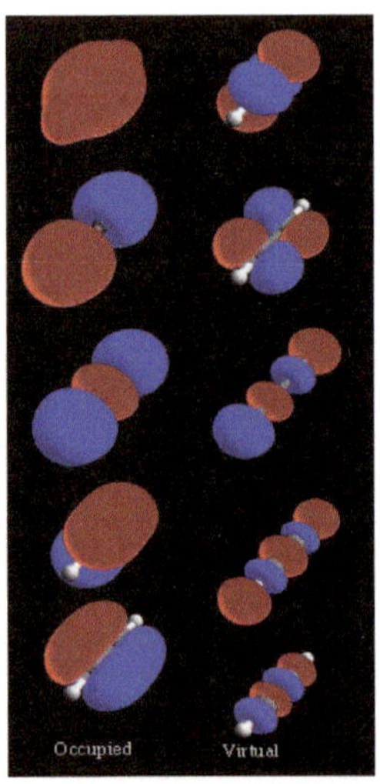

Complete acetylene (H−C≡C−H) molecular orbital set. The left column shows MO's which are occupied in the ground state, with the lowest-energy orbital at the top. The white and grey line visible in some MO's is the molecular axis passing through the nuclei. The orbital wave functions are positive in the red regions and negative in the blue. The right column shows virtual MO's which are empty in the ground state, but may be occupied in excited states.

In chemistry, a molecular orbital (MO) is a mathematical function describing the wave-like behavior of an electron in a molecule. This function can be used to calculate chemical and physical properties such as the probability of finding an electron in any specific region. The term orbital was introduced by Robert S. Mulliken in 1932 as an abbreviation for one-electron orbital wave function. At an elementary level, it is used to describe the region of space in which the function has a significant amplitude. Molecular orbitals are usually constructed by combining atomic orbitals or hybrid orbitals from each atom of the molecule, or other molecular orbitals from groups of atoms. They can be quantitatively calculated using the Hartree–Fock or self-consistent field (SCF) methods.

Overview

A molecular orbital (MO) can be used to represent the regions in a molecule where an electron occupying that orbital is likely to be found. Molecular orbitals are obtained from the combination of atomic orbitals, which predict the location of an electron in an atom. A molecular orbital can specify the electron configuration of a molecule: the spatial distribution and energy of one (or one pair of) electron(s). Most commonly a MO is represented as a linear combination of atomic orbitals (the LCAO-MO method), especially in qualitative or very approximate usage. They are invaluable in providing a simple model of bonding in molecules, understood through molecular orbital theory. Most present-day methods in computational chemistry begin by calculating the MOs of the system. A molecular orbital describes the behavior of one electron in the electric field generated by the nuclei and some average distribution of the other electrons. In the case of two electrons occupying the same orbital, the Pauli principle demands that they have opposite spin. Necessarily this is an approximation, and highly accurate descriptions of the molecular electronic wave function do not have orbitals.

Formation of Molecular Orbitals

Molecular orbitals arise from allowed interactions between atomic orbitals, which are allowed if the symmetries (determined from group theory) of the atomic orbitals are compatible with each other. Efficiency of atomic orbital interactions is determined from the overlap (a measure of how well two orbitals constructively interact with one another) between two atomic orbitals, which is significant if the atomic orbitals are close in energy. Finally, the number of molecular orbitals that form must equal the number of atomic orbitals in the atoms being combined to form the molecule.

Qualitative Discussion

For an imprecise, but qualitatively useful, discussion of the molecular structure, the molecular orbitals can be obtained from the "Linear combination of atomic orbitals molecular orbital method" ansatz. Here, the molecular orbitals are expressed as linear combinations of atomic orbitals.

Linear Combinations of Atomic Orbitals (LCAO)

Molecular orbitals were first introduced by Friedrich Hund and Robert S. Mulliken in 1927 and 1928. The linear combination of atomic orbitals or "LCAO" approximation for molecular orbitals was introduced in 1929 by Sir John Lennard-Jones. His ground-breaking paper showed how to derive the electronic structure of the fluorine and oxygen molecules from quantum principles. This qualitative approach to molecular orbital theory is part of the start of modern quantum chemistry. Linear combinations of atomic orbitals (LCAO) can be used to estimate the molecular orbitals that are formed upon bonding between the molecule's constituent atoms. Similar to an atomic orbital, a Schrödinger equation, which describes the behavior of an electron, can be constructed for a molecular orbital as well. Linear combinations of atomic orbitals, or the sums and differences of the atomic wavefunctions, provide approximate solutions to the Hartree–Fock equations which correspond to the independent-particle approximation of the molecular Schrödinger equation. For simple diatomic molecules, the wavefunctions obtained are represented mathematically by the equations

$$\Psi = c_a\psi_a + c_b\psi_b$$

$$\Psi^* = c_a\psi_a - c_b\psi_b$$

where Ψ and Ψ^* are the molecular wavefunctions for the bonding and antibonding molecular orbitals, respectively, ψ_a and ψ_b are the atomic wavefunctions from atoms a and b, respectively, and c_a and c_b are adjustable coefficients. These coefficients can be positive or negative, depending on the energies and symmetries of the individual atomic orbitals. As the two atoms become closer together, their atomic orbitals overlap to produce areas of high electron density, and, as a consequence, molecular orbitals are formed between the two atoms. The atoms are held together by the electrostatic attraction between the positively charged nuclei and the negatively charged electrons occupying bonding molecular orbitals.

Bonding, Antibonding, and Nonbonding MOs

When atomic orbitals interact, the resulting molecular orbital can be of three types: bonding, antibonding, or nonbonding.

Bonding MOs:

- Bonding interactions between atomic orbitals are constructive (in-phase) interactions.
- Bonding MOs are lower in energy than the atomic orbitals that combine to produce them.

Antibonding MOs:

- Antibonding interactions between atomic orbitals are destructive (out-of-phase) interactions, with a nodal plane where the wavefunction of the antibonding orbital is zero between the two interacting atoms
- Antibonding MOs are higher in energy than the atomic orbitals that combine to produce them.

Nonbonding MOs:

- Nonbonding MOs are the result of no interaction between atomic orbitals because of lack of compatible symmetries.
- Nonbonding MOs will have the same energy as the atomic orbitals of one of the atoms in the molecule.

Sigma and Pi labels for MOs

The type of interaction between atomic orbitals can be further categorized by the molecular-orbital symmetry labels σ (sigma), π (pi), δ (delta), φ (phi), γ (gamma) etc. paralleling the symmetry of the atomic orbitals s, p, d, f and g. The number of nodal planes containing the internuclear axis between the atoms concerned is zero for σ MOs, one for π, two for δ, etc.

σ Symmetry

A MO with σ symmetry results from the interaction of either two atomic s-orbitals or two atomic p_z-orbitals. An MO will have σ-symmetry if the orbital is symmetric with respect to the axis joining the two nuclear centers, the internuclear axis. This means that rotation of the MO about the internuclear axis does not result in a phase change. A σ* orbital, sigma antibonding orbital, also maintains the same phase when rotated about the internuclear axis. The σ* orbital has a nodal plane that is between the nuclei and perpendicular to the internuclear axis.

π Symmetry

A MO with π symmetry results from the interaction of either two atomic p_x orbitals or p_y orbitals. An MO will have π symmetry if the orbital is asymmetric with respect to rotation about the internuclear axis. This means that rotation of the MO about the internuclear axis will result in a phase change. There is one nodal plane containing the internuclear axis, if real orbitals are considered.

A π* orbital, pi antibonding orbital, will also produce a phase change when rotated about the internuclear axis. The π* orbital also has a second nodal plane between the nuclei.

δ Symmetry

A MO with δ symmetry results from the interaction of two atomic d_{xy} or $d_{x^2-y^2}$ orbitals. Because these molecular orbitals involve low-energy d atomic orbitals, they are seen in transition-metal complexes. A δ bonding orbital has two nodal planes containing the internuclear axis, and a δ* antibonding orbital also has a third nodal plane between the nuclei.

φ Symmetry

Suitably aligned f atomic orbitals overlap to form phi molecular orbital (a phi bond)

Theoretical chemists have conjectured that higher-order bonds, such as phi bonds corresponding to overlap of f atomic orbitals, are possible. There is as of 2005 only one known example of a molecule purported to contain a phi bond (a U–U bond, in the molecule U_2).

Gerade and Ungerade Symmetry

For molecules that possess a center of inversion (centrosymmetric molecules) there are additional labels of symmetry that can be applied to molecular orbitals. Centrosymmetric molecules include:

- Homonuclear diatomics, X_2
- Octahedral, EX6
- Square planar, EX4.

Non-centrosymmetric molecules include:

- Heteronuclear diatomics, XY
- Tetrahedral, EX_4.

If inversion through the center of symmetry in a molecule results in the same phases for the molecular orbital, then the MO is said to have gerade (g) symmetry, from the German word for even. If inversion through the center of symmetry in a molecule results in a phase change for the molecular orbital, then the MO is said to have ungerade (u) symmetry, from the German word for odd. For a bonding MO with σ-symmetry, the orbital is σ_g (s' + s" is symmetric), while an antibonding MO with σ-symmetry the orbital is σ_u, because inversion of s' – s" is antisymmetric. For a bonding MO with π-symmetry the orbital is π_u because inversion through the center of symmetry for would produce a sign change (the two p atomic orbitals are in phase with each other but the two lobes have opposite signs), while an antibonding MO with π-symmetry is π_g because inversion through the center of symmetry for would not produce a sign change (the two p orbitals are antisymmetric by phase).

MO Diagrams

The qualitative approach of MO analysis uses a molecular orbital diagram to visualize bonding interactions in a molecule. In this type of diagram, the molecular orbitals are represented by horizontal lines; the higher a line the higher the energy of the orbital, and degenerate orbitals are placed on the same level with a space between them. Then, the electrons to be placed in the molecular orbitals are slotted in one by one, keeping in mind the Pauli exclusion principle and Hund's rule of maximum multiplicity (only 2 electrons, having opposite spins, per orbital; place as many unpaired electrons on one energy level as possible before starting to pair them). For more complicated molecules, the wave mechanics approach loses utility in a qualitative understanding of bonding (although is still necessary for a quantitative approach). Some properties:

- A basis set of orbitals includes those atomic orbitals that are available for molecular orbital interactions, which may be bonding or antibonding
- The number of molecular orbitals is equal to the number of atomic orbitals included in the linear expansion or the basis set
- If the molecule has some symmetry, the degenerate atomic orbitals (with the same atomic energy) are grouped in linear combinations (called symmetry-adapted atomic orbitals

(SO)), which belong to the representation of the symmetry group, so the wave functions that describe the group are known as symmetry-adapted linear combinations (SALC).

- The number of molecular orbitals belonging to one group representation is equal to the number of symmetry-adapted atomic orbitals belonging to this representation
- Within a particular representation, the symmetry-adapted atomic orbitals mix more if their atomic energy levels are closer.

The general procedure for constructing a molecular orbital diagram for a reasonably simple molecule can be summarized as follows:

1. Assign a point group to the molecule.

2. Look up the shapes of the SALCs.

3. Arrange the SALCs of each molecular fragment in increasing order of energy, first noting whether they stem from *s*, *p*, or *d* orbitals (and put them in the order $s < p < d$), and then their number of internuclear nodes.

4. Combine SALCs of the same symmetry type from the two fragments, and from N SALCs form N molecular orbitals.

5. Estimate the relative energies of the molecular orbitals from considerations of overlap and relative energies of the parent orbitals, and draw the levels on a molecular orbital energy level diagram (showing the origin of the orbitals).

6. Confirm, correct, and revise this qualitative order by carrying out a molecular orbital calculation by using commercial software.

Bonding in Molecular Orbitals

Orbital Degeneracy

Molecular orbitals are said to be degenerate if they have the same energy. For example, in the homonuclear diatomic molecules of the first ten elements, the molecular orbitals derived from the p_x and the p_y atomic orbitals result in two degenerate bonding orbitals (of low energy) and two degenerate antibonding orbitals (of high energy).

Ionic Bonds

When the energy difference between the atomic orbitals of two atoms is quite large, one atom's orbitals contribute almost entirely to the bonding orbitals, and the others atom's orbitals contribute almost entirely to the antibonding orbitals. Thus, the situation is effectively that one or more electrons have been transferred from one atom to the other. This is called an (mostly) ionic bond.

Bond Order

The bond order, or number of bonds, of a molecule can be determined by combining the number of electrons in bonding and antibonding molecular orbitals. A pair of electrons in a bonding orbital creates a bond, whereas a pair of electrons in an antibonding orbital negates a bond. For example, N_2, with eight electrons in bonding orbitals and two electrons in antibonding orbitals, has a bond order of three, which constitutes a triple bond.

Bond strength is proportional to bond order—a greater amount of bonding produces a more stable bond—and bond length is inversely proportional to it—a stronger bond is shorter.

There are rare exceptions to the requirement of molecule having a positive bond order. Although Be_2 has a bond order of 0 according to MO analysis, there is experimental evidence of a highly unstable Be_2 molecule having a bond length of 245 pm and bond energy of 10 kJ/mol.

HOMO and LUMO

The highest occupied molecular orbital and lowest unoccupied molecular orbital are often referred to as the HOMO and LUMO, respectively. The difference of the energies of the HOMO and LUMO, termed the band gap, can sometimes serve as a measure of the excitability of the molecule: The smaller the energy the more easily it will be excited.

Molecular Orbital Examples

Homonuclear Diatomics

Homonuclear diatomic MOs contain equal contributions from each atomic orbital in the basis set. This is shown in the homonuclear diatomic MO diagrams for H_2, He_2, and Li_2, all of which containing symmetric orbitals.

H_2

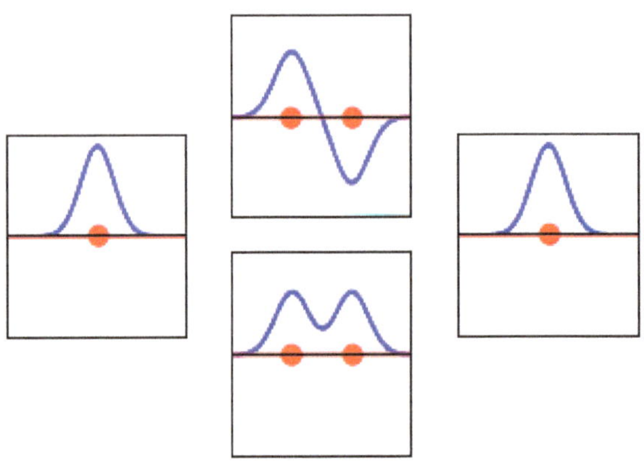

Electron wavefunctions for the 1s orbital of a lone hydrogen atom (left and right) and the corresponding bonding (bottom) and antibonding (top) molecular orbitals of the H_2 molecule. The real part of the wavefunction is the blue curve, and the imaginary part is the red curve. The red dots mark the locations of the nuclei. The electron wavefunction oscillates according to the Schrödinger wave equation, and orbitals are its standing waves. The standing wave frequency is proportional to the orbital's kinetic energy. (This plot is a one-dimensional slice through the three-dimensional system.)

As a simple MO example consider the hydrogen molecule, H_2, with the two atoms labelled H' and

H". The lowest-energy atomic orbitals, 1s' and 1s", do not transform according to the symmetries of the molecule. However, the following symmetry adapted atomic orbitals do:

1s' – 1s"	Antisymmetric combination: negated by reflection, unchanged by other operations
1s' + 1s"	Symmetric combination: unchanged by all symmetry operations

The symmetric combination (called a bonding orbital) is lower in energy than the basis orbitals, and the antisymmetric combination (called an antibonding orbital) is higher. Because the H_2 molecule has two electrons, they can both go in the bonding orbital, making the system lower in energy (and, hence, more stable) than two free hydrogen atoms. This is called a covalent bond. The bond order is equal to the number of bonding electrons minus the number of antibonding electrons, divided by 2. In this example, there are 2 electrons in the bonding orbital and none in the antibonding orbital; the bond order is 1, and there is a single bond between the two hydrogen atoms.

He_2

On the other hand, consider the hypothetical molecule of He_2 with the atoms labeled He' and He". As with H_2, the lowest-energy atomic orbitals are the 1s' and 1s", and do not transform according to the symmetries of the molecule, while the symmetry adapted atomic orbitals do. The symmetric combination—the bonding orbital—is lower in energy than the basis orbitals, and the antisymmetric combination—the antibonding orbital—is higher. Unlike H_2, with two valence electrons, He_2 has four in its neutral ground state. Two electrons fill the lower-energy bonding orbital, $\sigma_g(1s)$, while the remaining two fill the higher-energy antibonding orbital, $\sigma_u^*(1s)$. Thus, the resulting electron density around the molecule does not support the formation of a bond between the two atoms; without a stable bond holding the atoms together, molecule would not be expected to exist. Another way of looking at it is that there are two bonding electrons and two antibonding electrons; therefore, the bond order is 0 and no bond exists (the molecule has one bound state supported by the Van der Waals potential).

Li_2

Dilithium Li_2 is formed from the overlap of the 1s and 2s atomic orbitals (the basis set) of two Li atoms. Each Li atom contributes three electrons for bonding interactions, and the six electrons fill the three MOs of lowest energy, $\sigma_g(1s)$, $\sigma_u^*(1s)$, and $\sigma_g(2s)$. Using the equation for bond order, it is found that dilithium has a bond order of one, a single bond.

Noble Gases

Considering a hypothetical molecule of He_2, since the basis set of atomic orbitals is the same as in the case of H_2, we find that both the bonding and antibonding orbitals are filled, so there is no energy advantage to the pair. HeH would have a slight energy advantage, but not as much as H_2 + 2 He, so the molecule is very unstable and exists only briefly before decomposing into hydrogen and helium. In general, we find that atoms such as He that have full energy shells rarely bond with other atoms. Except for short-lived Van der Waals complexes, there are very few noble gas compounds known.

Heteronuclear Diatomics

While MOs for homonuclear diatomic molecules contain equal contributions from each interacting atomic orbital, MOs for heteronuclear diatomics contain different atomic orbital contributions. Orbital interactions to produce bonding or antibonding orbitals in heteronuclear diatomics occur if there is sufficient overlap between atomic orbitals as determined by their symmetries and similarity in orbital energies.

HF

In hydrogen fluoride HF overlap between the H 1s and F 2s orbitals is allowed by symmetry but the difference in energy between the two atomic orbitals prevents them from interacting to create a molecular orbital. Overlap between the H 1s and F $2p_z$ orbitals is also symmetry allowed, and these two atomic orbitals have a small energy separation. Thus, they interact, leading to creation of σ and σ* MOs and a molecule with a bond order of 1. Since HF is a non-centrosymmetric molecule, the symmetry labels g and u do not apply to its molecular orbitals.

Quantitative Approach

To obtain quantitative values for the molecular energy levels, one needs to have molecular orbitals that are such that the configuration interaction (CI) expansion converges fast towards the full CI limit. The most common method to obtain such functions is the Hartree–Fock method, which expresses the molecular orbitals as eigenfunctions of the Fock operator. One usually solves this problem by expanding the molecular orbitals as linear combinations of Gaussian functions centered on the atomic nuclei. The equation for the coefficients of these linear combinations is a generalized eigenvalue equation known as the Roothaan equations, which are in fact a particular representation of the Hartree–Fock equation. There are a number of programs in which quantum chemical calculations of MOs can be performed, including Spartan and HyperChem.

Simple accounts often suggest that experimental molecular orbital energies can be obtained by the methods of ultra-violet photoelectron spectroscopy for valence orbitals and X-ray photoelectron spectroscopy for core orbitals. This, however, is incorrect as these experiments measure the ionization energy, the difference in energy between the molecule and one of the ions resulting from the removal of one electron. Ionization energies are linked approximately to orbital energies by Koopmans' theorem. While the agreement between these two values can be close for some molecules, it can be very poor in other cases.

Molecular orbitals are formed from the combinations of atomic orbitals.

Since orbitals are wavefunctions, they can combine either constructively to form bonding molecular orbitals or destructively to form antibonding molecular orbitals.

Why do We Need Molecular Orbital Theory?

Valence bond theory generally fails to explain the bonding in simple molecules. On the other hand, molecular orbital theory is better approach for the molecules those are having extended p systems. With the help of molecular orbital one can understand the electronic transitions in molecules.

H2+ -ion is stable because of the bonding between the two hydrogen atoms which form a bound state with one electron.

Similar to this, if we consider two hydrogen atoms to form hydrogen molecule, we expect that the wavefunctions of two hydrogen atoms $\phi(a)$ and $\phi(b)$ will interact constructively or destructively.

If they interact constructively then the molecular wavefunction

$$\psi_+ = \frac{1}{\sqrt{2}}(\phi_{1s}(a) + \phi_{1s}(b))$$

This will form bonding orbital. And if they interact destructively, it will form antibonding orbital

$$\psi_- = \frac{1}{\sqrt{2}}(\phi_{1s}(a) - \phi_{1s}(b))$$

We have also realized that the degenerate levels of the two hydrogen atoms will be separated as the two nuclei come closer to each other.

As shown in figure-a, the ground state of two hydrogen atomic orbitals split into two energy levels – bonding ψ_+ and anitibonding ψ_-

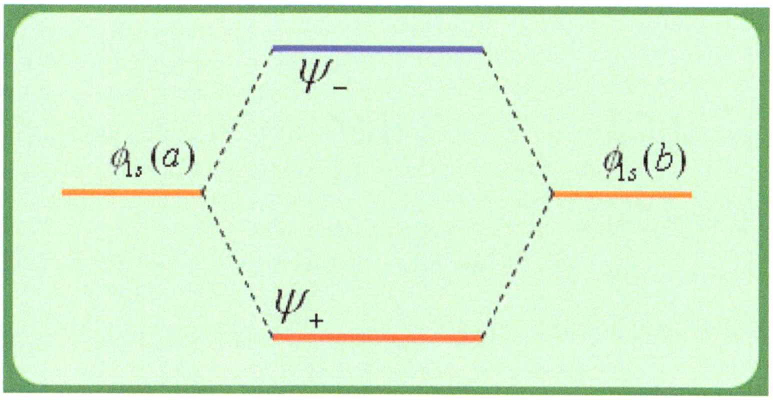

Figure (a)

Figure-b explains the energies of bonding and anti-bonding molecular orbitals for first row di-atomic molecules.

Two electrons in H_2 occupy bonding molecular orbital, with anti-parallel spins. If irradiated by UV light, molecule may absorb energy and promote one electron into its anti-bonding orbital.

The filling of lower molecular orbital indicates that the molecule is stable compared to two individual atoms.

"+" and "-" sign indicates the sign of the wavefunction.

σ_{1s} is symmetric with respect to the center of inversion (middle point of the bond), this sometimes called as symmetric or gerade and designate as σ_{1sg}

σ^*_{1s} is not symmetric with respect to the center of inversion (middle point of the bond), this sometimes called as asymmetric or ungerade and designate as $\sigma^*_{1s\mu}$

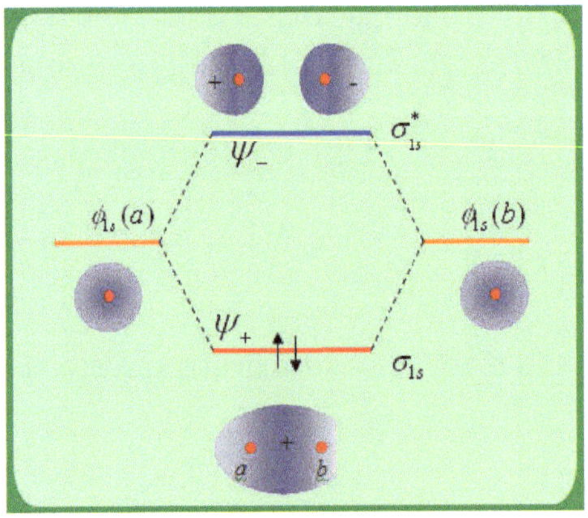

Figure (b)

Molecular orbital form when atomic orbitals with similar energies and proper symmetry spatially overlap each other.

For homonuclear molecule such as H_2, N_2, ...etc.

s-orbitals combine with *s*-orbitals and *p*-orbitals combine with *p*-orbitals. In figure we have seen the interaction of *s* orbital with another *s* orbital.

For p- orbitals :

There are three *px*, *py* and *pz* orbitals. If we consider X-axis coincides with internuclear axis then for *px* orbitals, there will be end-on-end overlap along the bond axis as shown in figure-c. It forms s bonding and nonbonding orbitals.

For s bonding orbital, it is symmetric about the center of inversion and thus it is σ_{2yg}

For s antibonding orbital, it is asymmetric about the center of inversion and thus it is $\sigma^*_{2p\mu}$

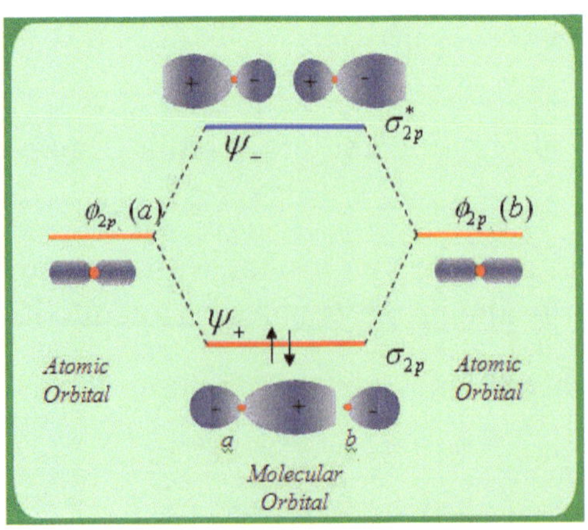

Figure (c)

The other *px*, and *py* orbitals overlap side-by-side. This is known as p-bonding. This overlap is less than the overlap along the bond axis.

Note that, the overlap is more ® it will be more stable and thus energy will be lower.

Thus, the p-bonding orbital will be higher in energy than that of the s-orbital.

p-orbitals are asymmetric with respect to the bond axis as shown in figure. No electron density surrounding the bond axis. It has node along the internuclear axis.

is asymmetric about the center of inversion, and thus $\pi_{2p,u}$

π^*_{2p} is symmetric about the center of inversion, and thus $\pi^*_{2p,g}$

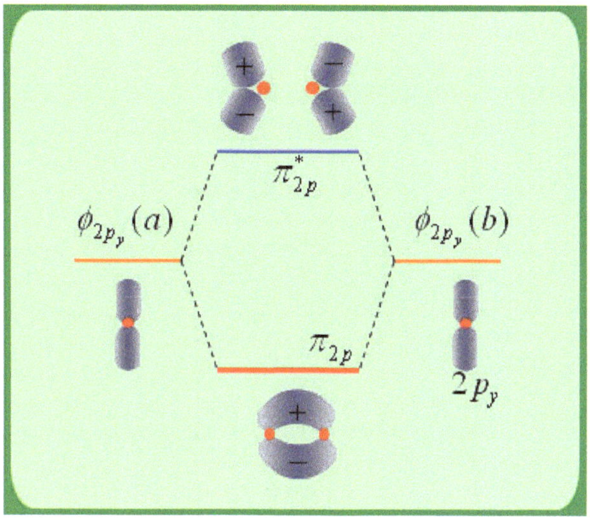

Figure (d)

Figure-e shows the total energy level diagram for p-orbitals interaction.

Note that s-orbital is the lowest in energy due to maximum overlap.

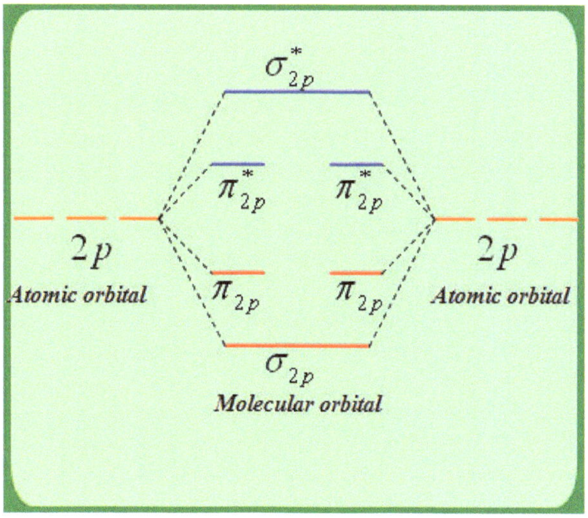

Figure (e)

Note:

Electron can occupy molecular orbital that are lower in energy.

Molecular orbitals may be empty or contain one or two electrons.

For two electrons occupied orbitals, electrons must be spin paired due to Pauli exclusion principle.

When occupying molecular orbitals, electrons occupy separate orbitals with parallel spin before pairing.

The highest occupied molecular orbital is known as HOMO.

The lowest unoccupied molecular orbital is known as LUMO.

Let us take an example of oxygen molecule.

Oxygen atom is having electron configuration $1s2^2s2^2p^4$. So the oxygen molecule is having total of 16 electrons.

$$1s^2.1s^2 \Rightarrow \sigma_{1s} \rightarrow \sigma_g$$
$$\Rightarrow \sigma^*_{1s} \rightarrow \sigma^*_\pi$$

$$2s^2, 2s^2 \Rightarrow \sigma_{2s} \rightarrow \sigma_s$$
$$\Rightarrow \sigma^*_{1s} \rightarrow \sigma^*_s$$

$$2p^2, 2p^2 \Rightarrow \sigma_{2p} \rightarrow \sigma_g$$
$$\Rightarrow \pi_{2p} \rightarrow \pi_y$$
$$\Rightarrow \pi_{2p} \rightarrow \pi^*_F$$
$$\Rightarrow \sigma^*_{2p} \rightarrow \sigma^*_\pi$$

The figure-f shows that for oxygen molecule, there are two unpaired electrons.

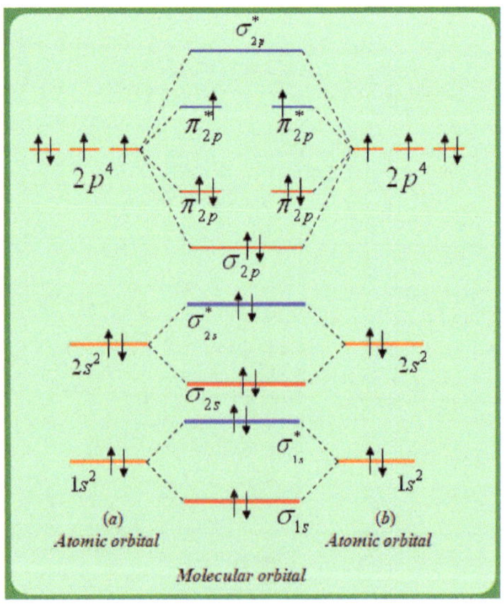

Figure (f)

Bond order indicates the strength of a bond and is defined as

Bond order = (no. of electrons in bonding orbitals)/2

– (no. of electrons in the antibonding orbitals)/2

For oxygen, considering only the valence electrons (2p orbital)

Bond order = (8-4)/2 = 2. So the result is a double bond.

Bond order =1 (single bond); Bond order = 2 (double bond) e.t.c

For heteronuclear atoms, due to the small energy gap between *s* and *p* orbitals, *s* and *p* orbitals overlap is possible. Figure-g shows the overlap between *s* and *p* orbital.

This interaction could be constructive or destructive.

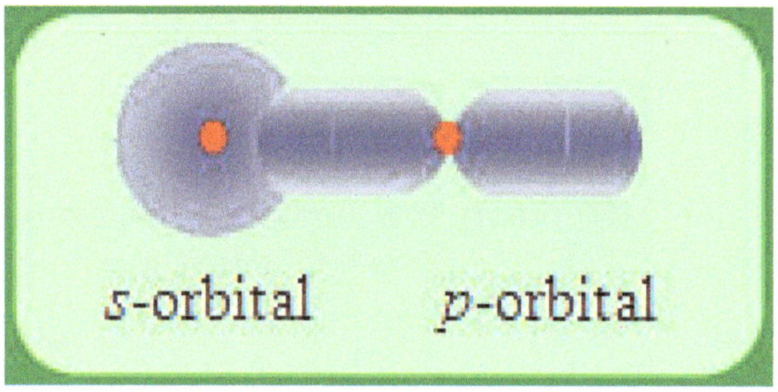

Figure (g)

Let us take an example of HF molecule. Figure-h shows the energy level diagram of the HF molecule. Hydrogen 1*s* orbital interact with 2*s* and 2*px* orbitals of fluorine. 2*py* and 2*pz* orbitals do not have the proper symmetry to interact with 1*s* orbital. So these are nonbonding orbital.

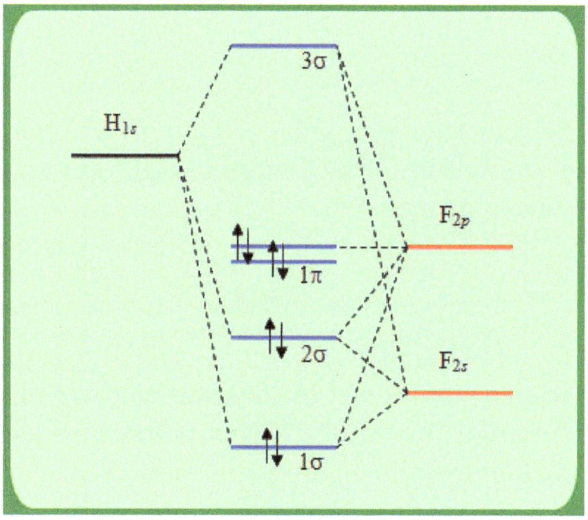

Figure (h)

Let us take another example of CO as shown in figure-i. Generally to simplify the things, interactions of the orbitals containg valence electrons are considered to form the molecular orbitals. The molecular orbital 2s is the highest filled orbital and is HOMO The molecular orbital 3s is the lowest unfilled orbital and is LUMO.

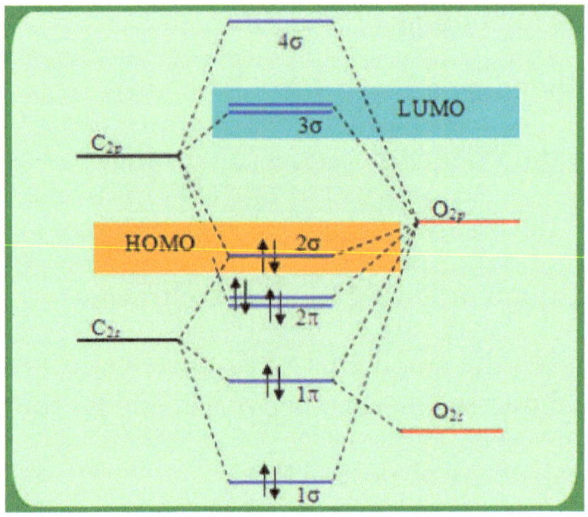

Figure (i)

Molecular Orbital Theory

In chemistry, molecular orbital (MO) theory is a method for determining molecular structure in which electrons are not assigned to individual bonds between atoms, but are treated as moving under the influence of the nuclei in the whole molecule. The spatial and energetic properties of electrons within atoms are fixed by quantum mechanics to form orbitals that contain these electrons. While atomic orbitals contain electrons ascribed to a single atom, molecular orbitals, which surround a number of atoms in a molecule, contain valence electrons between atoms. Molecular orbital theory, which was proposed in the early twentieth century, revolutionized the study of bonding by approximating the positions of bonded electrons—the molecular orbitals—as linear combinations of atomic orbitals (LCAO). These approximations are now made by applying the density functional theory (DFT) or Hartree–Fock (HF) models to the Schrödinger equation.

Quantitative Applications

In this theory, each molecule has a set of molecular orbitals, in which it is assumed that the molecular orbital wave function ψ_j can be written as a simple weighted sum of the n constituent atomic orbitals χ_i, according to the following equation:

$$\psi_j = \sum_{i=1}^{n} c_{ij}\chi_i.$$

One may determine c_{ij} coefficients numerically by substituting this equation into the Schrödinger's equation and applying the variational principle. The variational principle is a mathematical technique used in quantum mechanics to build up the coefficients of each atomic orbital basis. A larger coefficient means that the orbital basis is composed more of that particular contributing atomic orbital—hence, the molecular orbital is best characterized by that type. This method of quantifying orbital contribution as Linear Combinations of Atomic Orbitals is used in computational chemistry. An additional unitary transformation can be applied on the system to accelerate the convergence in some computational schemes. Molecular orbital theory was seen as a competitor to valence bond theory in the 1930s, before it was realized that the two methods are closely related and that when extended they become equivalent.

History

Molecular orbital theory was developed, in the years after valence bond theory had been established (1927), primarily through the efforts of Friedrich Hund, Robert Mulliken, John C. Slater, and John Lennard-Jones. MO theory was originally called the Hund-Mulliken theory. According to German physicist and physical chemist Erich Hückel, the first quantitative use of molecular orbital theory was the 1929 paper of Lennard-Jones. This paper notably predicted a triplet ground state for the dioxygen molecule which explained its paramagnetism before valence bond theory, which came up with its own explanation in 1931. The word *orbital* was introduced by Mulliken in 1932. By 1933, the molecular orbital theory had been accepted as a valid and useful theory.

Erich Hückel applied molecular orbital theory to unsaturated hydrocarbon molecules starting in 1931 with his Hückel molecular orbital (HMO) method for the determination of MO energies for pi electrons, which he applied to conjugated and aromatic hydrocarbons. This method provided an explanation of the stability of molecules with six pi-electrons such as benzene.

The first accurate calculation of a molecular orbital wavefunction was that made by Charles Coulson in 1938 on the hydrogen molecule. By 1950, molecular orbitals were completely defined as eigenfunctions (wave functions) of the self-consistent field Hamiltonian and it was at this point that molecular orbital theory became fully rigorous and consistent. This rigorous approach is known as the Hartree–Fock method for molecules although it had its origins in calculations on atoms. In calculations on molecules, the molecular orbitals are expanded in terms of an atomic orbital basis set, leading to the Roothaan equations. This led to the development of many ab initio quantum chemistry methods. In parallel, molecular orbital theory was applied in a more approximate manner using some empirically derived parameters in methods now known as semi-empirical quantum chemistry methods.

The success of Molecular Orbital Theory also spawned ligand field theory, which was developed during the 1930s and 1940s as an alternative to crystal field theory.

Types of Orbitals

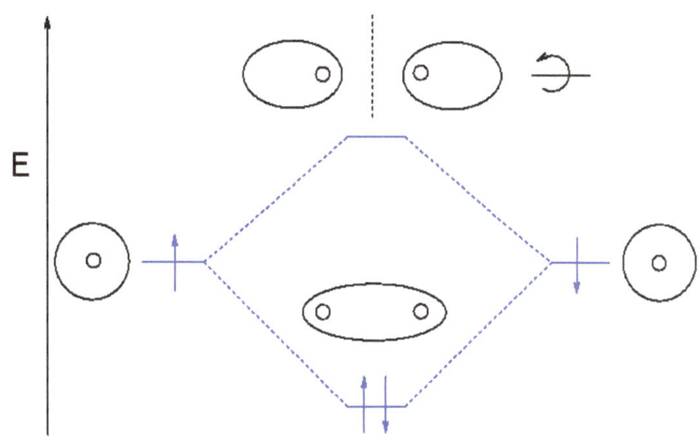

MO diagram showing the formation of molecular orbitals of H_2 (centre) from atomic orbitals of two H atoms. The lower-energy MO is bonding with electron density concentrated between the two H nuclei. The higher-energy MO is anti-bonding with electron density concentrated behind each H nucleus.

Molecular orbital (MO) theory uses a linear combination of atomic orbitals (LCAO) to represent molecular orbitals resulting from bonds between atoms. These are often divided into *bonding* orbitals, anti-bonding orbitals, and non-bonding orbitals. A bonding orbital concentrates electron density in the region *between* a given pair of atoms, so that its electron density will tend to attract each of the two nuclei toward the other and hold the two atoms together. An anti-bonding orbital concentrates electron density "behind" each nucleus (i.e. on the side of each atom which is farthest from the other atom), and so tends to pull each of the two nuclei away from the other and actually weaken the bond between the two nuclei. Electrons in non-bonding orbitals tend to be associated with atomic orbitals that do not interact positively or negatively with one another, and electrons in these orbitals neither contribute to nor detract from bond strength.

Molecular orbitals are further divided according to the types of atomic orbitals they are formed from. Chemical substances will form bonding interactions if their orbitals become lower in energy when they interact with each other. Different bonding orbitals are distinguished that differ by electron configuration (electron cloud shape) and by energy levels.

The molecular orbitals of a molecule can be illustrated in molecular orbital diagrams.

Overview

MO theory provides a global, delocalized perspective on chemical bonding. In MO theory, *any* electron in a molecule may be found *anywhere* in the molecule, since quantum conditions allow electrons to travel under the influence of an arbitrarily large number of nuclei, as long as they are in eigenstates permitted by certain quantum rules. Thus, when excited with the requisite amount of energy through high-frequency light or other means, electrons can transition to higher-energy molecular orbitals. For instance, in the simple case of a hydrogen diatomic molecule, promotion of a single electron from a bonding orbital to an antibonding orbital can occur under UV radiation. This promotion weakens the bond between the two hydrogen atoms and can lead to photodissociation—the breaking of a chemical bond due to the absorption of light.

Although in MO theory *some* molecular orbitals may hold electrons that are more localized between specific pairs of molecular atoms, *other* orbitals may hold electrons that are spread more uniformly over the molecule. Thus, overall, bonding is far more delocalized in MO theory, which makes it more applicable to resonant molecules that have equivalent non-integer bond orders than valence bond (VB) theory. This makes MO theory more useful for the description of extended systems.

An example is the MO description of benzene, C6H6, which is an aromatic hexagonal ring of six carbon atoms and three double bonds. In this molecule, 24 of the 30 total valence bonding electrons—24 coming from carbon atoms and 6 coming from hydrogen atoms—are located in 12 σ (sigma) bonding orbitals, which are located mostly between pairs of atoms (C-C or C-H), similarly to the electrons in the valence bond description. However, in benzene the remaining six bonding electrons are located in three π (pi) molecular bonding orbitals that are delocalized around the ring. Two of these electrons are in an MO that has equal orbital contributions from all six atoms. The other four electrons are in orbitals with vertical nodes at right angles to each other. As in the

VB theory, all of these six delocalized π electrons reside in a larger space that exists above and below the ring plane. All carbon-carbon bonds in benzene are chemically equivalent. In MO theory this is a direct consequence of the fact that the three molecular π orbitals combine and evenly spread the extra six electrons over six carbon atoms.

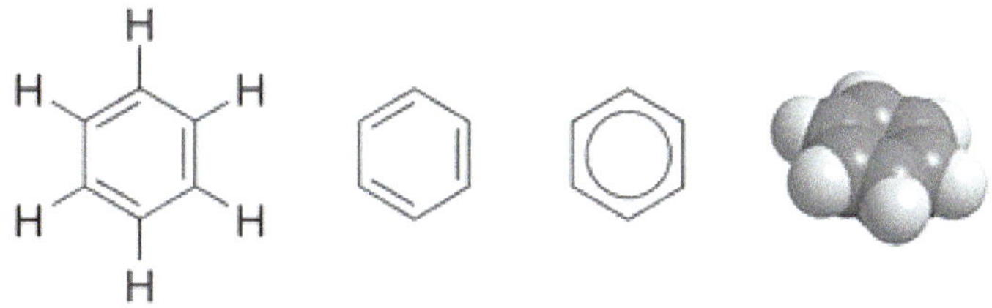

Structure of benzene

In molecules such as methane, CH_4, the eight valence electrons are found in four MOs that are spread out over all five atoms. However, it is possible to transform the MOs into four localized sp^3 orbitals. Linus Pauling, in 1931, hybridized the carbon 2s and 2p orbitals so that they pointed directly at the hydrogen 1s basis functions and featured maximal overlap. However, the delocalized MO description is more appropriate for predicting ionization energies and the positions of spectral absorption bands. When methane is ionized, a single electron is taken from the valence MOs, which can come from the s bonding or the triply degenerate p bonding levels, yielding two ionization energies. In comparison, the explanation in VB theory is more complicated. When one electron is removed from an sp^3 orbital, resonance is invoked between four valence bond structures, each of which has a single one-electron bond and three two-electron bonds. Triply degenerate T_2 and A_1 ionized states (CH_4^+) are produced from different linear combinations of these four structures. The difference in energy between the ionized and ground state gives the two ionization energies.

As in benzene, in substances such as beta carotene, chlorophyll, or heme, some electrons in the π orbitals are spread out in molecular orbitals over long distances in a molecule, resulting in light absorption in lower energies (the visible spectrum), which accounts for the characteristic colours of these substances. This and other spectroscopic data for molecules are well explained in MO theory, with an emphasis on electronic states associated with multicenter orbitals, including mixing of orbitals premised on principles of orbital symmetry matching. The same MO principles also naturally explain some electrical phenomena, such as high electrical conductivity in the planar direction of the hexagonal atomic sheets that exist in graphite. This results from continuous band overlap of half-filled p orbitals and explains electrical conduction. MO theory recognizes that some electrons in the graphite atomic sheets are completely delocalized over arbitrary distances, and reside in very large molecular orbitals that cover an entire graphite sheet, and some electrons are thus as free to move and therefore conduct electricity in the sheet plane, as if they resided in a metal.

References

- Gagliardi, Laura; Roos, Björn O. (2005). "Quantum chemical calculations show that the uranium molecule U2 has a quintuple bond". Nature. 433: 848–851. Bibcode:2005Natur.433..848G. doi:10.1038/nature03249. PMID 15729337
- Albright, T. A.; Burdett, J. K.; Whangbo, M.-H. (2013). Orbital Interactions in Chemistry. Hoboken, N.J.: Wiley. ISBN 9780471080398

- Bondybey, V.E. (1984). "Electronic structure and bonding of Be2". Chemical Physics Letters. 109 (5): 436–441. doi:10.1016/0009-2614(84)80339-5
- Atkins., Peter ... [et al]. (2006). Inorganic chemistry (4. ed.). New York: W.H. Freeman. p. 208. ISBN 978-0-7167-4878-6
- Hückel, Erich (1934). "Theory of free radicals of organic chemistry". Trans. Faraday Soc. 30: 40–52. doi:10.1039/TF9343000040
- Catherine E. Housecroft, Alan G, Sharpe, Inorganic Chemistry, Pearson Prentice Hall; 2nd Edition, 2005, ISBN 0130-39913-2, p. 41-43
- Lennard-Jones, J.E. (1929). "The electronic structure of some diatomic molecules". Trans. Faraday Soc. 25: 668–686. doi:10.1039/TF9292500668
- Pauling, Linus (1931). "The Nature of the Chemical Bond. II. The One-Electron Bond and the Three-Electron Bond.". J. Am. Chem. Soc. 53: 3225–3237. doi:10.1021/ja01360a004
- Hall, George G. Lennard-Jones Paper of 1929 "Foundations of Molecular Orbital Theory." Check |url= value (help). Advances in Quantum Chemistry. 22. Bibcode:1991AdQC...22....1H. doi:10.1016/S0065-3276(08)60361-5. ISBN 978-0-12-034822-0. ISSN 0065-3276
- Coulson, C.A. (1938), "Self-consistent field for molecular hydrogen", Mathematical Proceedings of the Cambridge Philosophical Society, 34 (2): 204–212, Bibcode:1938PCPS...34..204C, doi:10.1017/S0305004100020089

Permissions

All chapters in this book are published with permission under the Creative Commons Attribution Share Alike License or equivalent. Every chapter published in this book has been scrutinized by our experts. Their significance has been extensively debated. The topics covered herein carry significant information for a comprehensive understanding. They may even be implemented as practical applications or may be referred to as a beginning point for further studies.

We would like to thank the editorial team for lending their expertise to make the book truly unique. They have played a crucial role in the development of this book. Without their invaluable contributions this book wouldn't have been possible. They have made vital efforts to compile up to date information on the varied aspects of this subject to make this book a valuable addition to the collection of many professionals and students.

This book was conceptualized with the vision of imparting up-to-date and integrated information in this field. To ensure the same, a matchless editorial board was set up. Every individual on the board went through rigorous rounds of assessment to prove their worth. After which they invested a large part of their time researching and compiling the most relevant data for our readers.

The editorial board has been involved in producing this book since its inception. They have spent rigorous hours researching and exploring the diverse topics which have resulted in the successful publishing of this book. They have passed on their knowledge of decades through this book. To expedite this challenging task, the publisher supported the team at every step. A small team of assistant editors was also appointed to further simplify the editing procedure and attain best results for the readers.

Apart from the editorial board, the designing team has also invested a significant amount of their time in understanding the subject and creating the most relevant covers. They scrutinized every image to scout for the most suitable representation of the subject and create an appropriate cover for the book.

The publishing team has been an ardent support to the editorial, designing and production team. Their endless efforts to recruit the best for this project, has resulted in the accomplishment of this book. They are a veteran in the field of academics and their pool of knowledge is as vast as their experience in printing. Their expertise and guidance has proved useful at every step. Their uncompromising quality standards have made this book an exceptional effort. Their encouragement from time to time has been an inspiration for everyone.

The publisher and the editorial board hope that this book will prove to be a valuable piece of knowledge for students, practitioners and scholars across the globe.

Index